The Yearbook on History and
Interpretation of Phenomenology 2013

Contact Information
Yearbook on History and Interpretation of Phenomenology
Center for Phenomenological Studies
Department of Philosophy
Faculty of Philosophy and Arts
Trnava University
Hornopotočná 23
918 43 Trnava
Slovakia

PETER LANG
EDITION

The Yearbook on History and Interpretation of Phenomenology 2013

Person – Subject – Organism

An Overview of Interdisciplinary Insights

PETER LANG
EDITION

Bibliographic Information published by the Deutsche Nationalbibliothek
The Deutsche Nationalbibliothek lists this publication in the Deutsche Nationalbibliografie; detailed bibliographic data is available in the internet at http://dnb.d-nb.de.

The issue was produced at the Department of Philosophy of Trnava University (Slovakia) with financial support from the VEGA 1/0272/13 grant project "Interpersonality, Normativity, Generativity: Phenomenological Research" as well as from Trnava University.

ISSN 2198-8595
ISBN 978-3-631-65069-1 (Print)
E-ISBN 978-3-653-04480-5 (E-Book)
DOI 10.3726/978-3-653-04480-5

Peter Lang Edition is an Imprint of Peter Lang GmbH.

Peter Lang – Frankfurt am Main · Bern · Bruxelles · New York · Oxford · Warszawa · Wien

This book is part of the Peter Lang Edition list and was peer reviewed prior to publication.

www.peterlang.com

Contents

List of Abbreviations

The list of abbreviations used in the following contributions

Georges Canguilhem

CV	*La connaissance de la vie*
NP	*Le normal et le pathologique*

Gilles Deleuze

DR	*Différence et répétition*

Jacques Derrida

SP	*Speech and Phenomena*

Martin Heidegger

GA	*Gesamtausgabe*

Edmund Husserl

LI	*Logical Investigations*
FTL	*Formal and Transcendental Logic*
CM	*Cartesian Meditation*
EJ	*Experience and Judgment: Investigations in a Genealogy of Logic*
Hua	*Husserliana*

Immanuel Kant

KrV	*Kritik der reinen Vernunft*

Maurice Merleau-Ponty

PP	*Phénoménologie de la perception*
SC	*La structure du comportement*
VI	*Le visible et l'invisible*

Introduction

I would like to write a few words about the new yearbook, about its aims and why we have decided to publish it. The idea of *The Yearbook on History and Interpretation of Phenomenology* came from the Center for Phenomenological Studies which is part of the Department of Philosophy at Trnava University (Slovakia). The Center joins a group of researchers focusing on several aspects of phenomenology and its history, as well as phenomenological interpretations of various fields of science, art and religion. Some crucial questions have emerged in our cultural milieu of the Central Europe which interest us in our work and which we intend to discuss within the broader context of science.

The title of *The Yearbook* comprises two basic terms: History and Interpretation. We would like to publish contributions about the history of phenomenology because phenomenology has its own specific development anchored in the texts of Edmund Husserl, his predecessors and followers, its distinctive themes and problems set within the frame of the philosophical and scientific discussions of their period. We are also open to inquiries about the interpretation of phenomenology and to different approaches towards understanding phenomenological research, its systematic and methodological insights and its possible contributions to contemporary discussions both about pure philosophy and within the context of more interdisciplinary research. We are open to broad discussions with other philosophical schools of thought and are interested in addressing our common themes

These two purposes are, of course, intertwined. We are convinced that investigations into historical aspects of phenomenology, with its multiple shifts and essential changes, will be valuable not only for us, but also for our readers. In the same vein we believe that it is important to keep clarifying our understanding of various phenomenological motivations and keep returning to the key phenomenological questions. It may seem sometimes that phenomenology belongs to history and is just an artifact from the past. However, phenomenology still presents new and vivid challenges to scientific research; it still poses new problems animating scientific discussion about the limits of science or the crisis of institutions. It opens the door to specific phenomenological theories inspired not only by phenomenological resources, but also by many non-phenomenological approaches. At the same time, phenomenology itself is often be challenged by other scientific disciplines forcing it to rethink, re-clarify or modify its own approaches.

In short, this new annual journal has specific intentions and goals, and offers a space for contributors to create an integrated unit together with other phenomenological journals oriented more exclusively towards either Husserlian phenomenology or various applications of phenomenology. We appreciate broad research in contemporary phenomenology and often participate in such projects. However, our intention is to create our own subset of phenomenological research in the belief that this will yield fruitful results and open new possibilities both in phenomenology and other scientific disciplines.

The main topic of this current issue encompasses three areas of phenomenological research: person, subject, and organism. These three topics are interrelated in various ways. On the one hand, we address the question of Husserlian phenomenology of personhood and subjectivity, and on the other hand, we address a broader problem including epistemological, ontological and biological approaches. Those great traditional and contemporary themes of subjectivity and intersubjectivity, concept of person, community and interpersonality, questions of humanity, value and biological status of human beings all slowly became part of Edmund Husserl's focus. We intend to show that a number of inspiring and unexplored questions arose from these thematic areas, questions which are related to various specific and interconnected fields of study. It is up to the reader to decide whether we have been successful in this.

In conclusion, let me express our gratitude to Peter Lang Ltd. for including *The Yearbook* among its publishing activities. I would also like to thank all contributors, reviewers and editors. Our special thanks go to Jonathan Gresty for his help with proofreading this issue, to Aaron Fortune, and also to Prof. Dr. Dr. Georg Schuppener for his proofreading of German contribution. We are also grateful for the support of the VEGA grant-funded project no. 1/0272/13 by the Ministry of Education, Science, Research and Sport in Slovak Republic.

Anton Vydra

Operatively functioning subjectivity: Toward a history of the received interpretation

Elizabeth A. Behnke

Abstract

Husserl's notion of functioning subjectivity (functioning intentionality, functioning I, functioning Corporeality) was taken up by subsequent phenomenologists and transformed into a received interpretation still effective today. The present essay traces the trajectory of this historical development across works by Fink, Merleau-Ponty, Brand, Held, and Mohanty, concluding that the assumption of an operatively functioning anonymity at the heart of subjective life may be one of the motivations leading some phenomenologists to collaborate with the project of the naturalization of consciousness. I close by identifying some further investigations needed to pave the way for a comparison between Husserl's own rich notion of functioning subjectivity and the received interpretation.

Keywords: Husserl, functioning, operative, intentionality, anonymity

Husserl uses the term "functioning" *(Fungieren, Funktionieren)* as a noun, a verb, and an adjective (functioning subjectivity, functioning intentionality, functioning I, functioning Corporeality,[1] etc.); although he seldom offers anything like a definition of it, this notion plays a crucial role in many of his descriptions and analyses. Subsequently, other phenomenologists appropriated and transformed this notion, turning it into a technical term with a life of its own. The purpose of this essay is to sketch some major "landmarks" in the efficient history[2] of this transformation, beginning with a key passage from *Formal and Transcendental Logic*. My procedure, in other words, will be to offer a straightforward scholarly account, carried out within historical consciousness, of a chain of Husserl-interpretations pregiven to me in my own historical situation; the discussion not only of the scope and development of this notion in Husserl's own writings but also of related issues arising from the standpoint of

1 I will follow the convention of using capital letters to preserve the *Leib/Körper* distinction in English; thus Corporeality = *Leiblichkeit,* Body = *Leib,* etc.

2 I am taking the term "efficient history" *(Wirkungsgeschichte)* in the sense developed in Thomas Seebohm, *Hermeneutics: Method and Methodology* (Dordrecht: Kluwer Academic Publishers, 2004), §37 (cf. also §§35–36). My interest here is limited to identifying key moments in a historical trajectory and thus I shall not be providing detailed interpretations of each of the "landmark" texts considered.

a methodologically guided hermeneutical consciousness[3] must be reserved for other occasions.

Formal and Transcendental Logic (originally intended as an introduction to *Experience and Judgment*[4]) was written in a few months at the end of 1928 and beginning of 1929,[5] and was published in volume 10 of the *Jahrbuch für Philosophie und phänomenologische Forschung* (1929) as well as being printed separately. Although the notion of "functioning" can also be found elsewhere in this work, the initial moment of the trajectory I want to document is highlighted by Mohanty,[6] who suggests that this may be the first time the term "functioning" appears along with "living" *(lebendig)* in Husserl's writings (at least, we may add, in works Husserl published during his lifetime):

> The living intentionality carries me, predelineates, determines my entire practical comportment [...] even though as a living, functioning intentionality, it may be unthematic, undisclosed, and thus beyond my knowledge.[7]

Taken in its original context in §94 and in light of other references in the text to thematizing an intentional life that normally remains unthematic,[8] this sentence can be taken as a concise description of the ongoing style of naive experience in everyday life, in contrast to a full transcendental phenomenological description of the nexus of subjective capabilities and intentional achievements brought to light by constitutive analyses. In other words, it is the task of phenomenology to make constituted sense comprehensible by explicating the actual and potential performances of my own conscious life and its constituting intentionality (*Hua*

---

3 See Seebohm, *Hermeneutics*, esp. Part III. Note that methodical and methodological hermeneutics must be clearly distinguished from a Heideggerian philosophical hermeneutics.

4 See Landgrebe's introduction to Edmund Husserl, *Erfahrung und Urteil. Untersuchungen zur Genealogie der Logik,* ed. Ludwig Landgrebe (Prag: Academia-Verlag, 1939).

5 See *Hua* XVII, xvii, xxi ff.; cf. Dorion Cairns, *Conversations with Husserl and Fink,* ed. Husserl-Archives (The Hague: Martinus Nijhoff, 1976), 20, 60f., and Ronald Bruzina, "Translator's Introduction," in Eugen Fink, *Sixth Cartesian Meditation: The Idea of a Transcendental Theory of Method, with textual notations by Edmund Husserl* (Bloomington, IN: Indiana University Press, 1995), lxxiii, n. 45.

6 J. N. Mohanty, "Husserl's Concept of Intentionality," *Analecta Husserliana* 1 (1971): 114; *The Concept of Intentionality* (St. Louis: Warren H. Green, 1972), 122.

7 "Die lebendige Intentionalität trägt mich, zeichnet vor, bestimmt mich praktisch in meinem ganzen Verhalten, [...] mag sie auch als lebendig fungierende unthematisch, unenthüllt und somit meinem Wissen entzogen sein" (*Hua* XVII, 242). All translations in the present essay are my own.

8 See esp. *Hua* XVII, 38, 184f.; see also 17, 20, 44, 116, 181, 215f., and cf. 179, 194, 241, 278, 281, 285.

XVII, 241; cf. 249, 253f.). But in the natural attitude, no explication is needed—the world and its myriad things are simply "there" for me to be experienced, accepted as holding good, with this or that made thematic while the thematizing life itself remains hidden, unthematic, anonymous (*Hua* XVII, 38, 185). Thus in *Formal and Transcendental Logic* the notion of functioning subjectivity emerges through a contrast between experience as lived in mundane life and the transcendental experience of a phenomenological researcher who systematically inquires into the structures of the functioning intentionality sustaining mundane life (*Hua* XVII, 241). How are these matters initially taken up in the literature?

The first figure shaping what will become a particular way of understanding anonymous functioning was Eugen Fink, who was Husserl's assistant from 1928 to 1937.[9] A fuller treatment of this problematic would have to include Fink's own philosophical approach to such matters, preserved in his manuscripts and now the topic of a major study by Ronald Bruzina;[10] Fink's Sixth Cartesian Meditation (a text with a significant efficient history of its own) is also relevant.[11] Here, however, I will focus on his references to this notion in two essays published in the 1930s.[12] Fink's 1933 *Kant-Studien* article contrasts the natural attitude's directedness toward the object that is present for it with Husserl's

---

9 See Ronald Bruzina, *Edmund Husserl and Eugen Fink: Beginnings and Ends in Phenomenology, 1928–1938* (New Haven, CT: Yale University Press, 2004), Ch. 1.

10 *Ibid.* See Eugen Fink, *Phänomenologische Werkstatt,* ed. Ronald Bruzina, 4 vols. (Freiburg: Karl Alber, 2006ff.), for manuscript materials now available in published form. However, this material remained unknown for many years and did not enter into the historical arc I am tracing; here I shall merely refer the reader to Bruzina, *Edmund Husserl and Eugen Fink*, e.g., 179, 197ff., 280ff., 357f., 463f., 499ff., *et passim*, and cf. also Sebastian Luft, *"Phänomenologie der Phänomenologie." Systematik und Methodologie der Phänomenologie in der Auseinandersetzung zwischen Husserl und Fink* (Dordrecht: Kluwer Academic Publishers, 2002); Guy van Kerckhoven, *Mundanisierung und Individuation bei E. Husserl und E. Fink. Die VI. Cartesianische Meditation und ihr "Einsatz"* (Würzberg: Königshausen & Neumann, 2003).

11 See Eugen Fink, *VI. Cartesianische Meditation. Teil 1: Die Idee einer transzendentalen Methodenlehre,* ed. Hans Ebeling, Jann Holl, and Guy van Kerckhoven; *Teil 2: Ergänzungsband,* ed. Guy van Kerckhoven, *Husserliana Dokumente II* (Dordrecht: Kluwer Academic Publishers, 1988), and cf. Ronald Bruzina, "Die Notizen Eugen Finks zur Umarbeitung von Edmund Husserls 'Cartesianischen Meditationen,'" *Husserl Studies* 6 (1989): 97–128; Guy van Kerckhoven, "Die 'Idee' einer transzendentalen Methodenlehre in phänomenologischer Absicht. Eine Voranzeige zu Eugen Finks Entwurf einer VI. Meditation zu Husserls *Méditations cartésiennes*," in *Phänomenologie im Widerstreit. Zum 50. Todestag Edmund Husserls*, ed. Christoph Jamme and Otto Pöggeler (Frankfurt am Main: Suhrkamp, 1989), 81–105.

12 Fink's later article "Operative Begriffe in Husserls Phänomenologie," *Zeitschrift für philosophische Forschung* 11 (1957): 321–337, will be considered below.

method of intentional analysis, which lays bare "the entire functioning system of interpenetrating presenting and appresenting intentions, of interpenetrating anticipations and habitual acquisitions, etc."[13] as the undisclosed and "anonymous" basis upon which the object comes to givenness.[14] In this way a region is opened up for phenomenological work,[15] and the task is to thematize the living functioning of transcendental subjectivity[16]—a task rendered all the more difficult because constituting transcendental life is "concealed" by the psychic intentional life that is its mundanized form.[17] Although Fink does not cite the passage from *Formal and Transcendental Logic* quoted above, he does refer to the latter work in other connections, and his presentation of living, functioning intentionality is consistent with the passage in question. But his 1939 article on the problem of Husserl's phenomenology proved even more influential in disseminating and reformulating the question of functioning intentionality.[18] Here Fink is concerned to frame the thematization of subjective life in terms of the question of being,[19] and he addresses the "executive function" of consciousness in its experiential meeting with entities,[20] seeking to bring to light the hidden yet effective intentional performances contributing to the seemingly simple "acts" whose correlates are "objects."[21] The following passage summarizes his point and serves as a "landmark" along the trajectory of the efficient history of this notion:

---

13 Note that in this reference to "das ganze fungierende System des Ineinander von präsentierenden und appräsentierenden Intentionen, von Antizipationen und habituellen Erwerben usw."—Eugen Fink, *Studien zur Phänomenologie 1930–1939* (Den Haag: Martinus Nijhoff, 1966)—Fink uses Husserl's term *"Ineinander,"* a notion that would later prove important to Merleau-Ponty.

14 *Ibid.*, 91

15 *Ibid.*, 94; cf. 138.

16 *Ibid.*, 122.

17 *Ibid.*, 142f.

18 For instance, it not only influences Merleau-Ponty, but is cited in Gerd Brand, *Welt, Ich und Zeit. Nach unveröffentlichten Manuskripten Edmund Husserls* (Den Haag: Martinus Nijhoff, 1955), xv, 4, 23, 39, and in Mohanty, "Husserl's Concept of Intentionality," 115; *The Concept of Intentionality*, 122. In addition, Klaus Held gives credit to this work (along with Landgrebe's 1962 "Husserls Abschied vom Cartesianismus") for stimulating his own study—Klaus Held, *Lebendige Gegenwart. Die Frage nach der Seinsweise des transzendentalen Ich bei Edmund Husserl, entwickelt am Leitfaden der Zeitproblematik* (Den Haag: Martinus Nijhoff, 1966), xiii.

19 Fink, *Studien zur Phänomenologie 1930–1939*, 189, 201; cf. 192f. on the "forgottenness" of being and 203, 208, 212 on intentional references as obscure, latent, concealed, implicit.

20 *Ibid.*, 210, 212.

21 *Ibid.*, 218f.

> The task of intentional analysis is to uncover the hidden efficacy of the *sense-filled* modes of consciousness that are veiled by their own results. The theme of such analyses is *functioning intentionality,* the living sense-forming, sense-achieving, sense-transforming function of consciousness that conceals itself in being consolidated into the simple psychic unities of acts.[22]

Thus the functioning intentionalities are revealed by an analysis that takes the complexity of noematic horizons as a clue to inquire back into the syntheses of noetic performances lying latent within each act[23] as moments of conscious functioning that remain out of awareness, concealed by their own results, and are here revealed for the first time.[24] Fink concludes by asking whether the functioning sense-forming life can become an "object" in the usual sense of things or objects in general, since it is the living, functioning system of intentionalities *through which* "acts" have "objects" at all.[25] Now it is true that Husserl's 1929 passage acknowledges that the living intentionality functions "even though" it "may" be unthematic, undisclosed. But ten years later we find Fink emphasizing functioning intentionality as a "hidden ground"[26] that can be wrested from its concealment only by introducing a "constructive moment" into phenomenological method.[27] And this sense of "construction" as a way of disclosing what cannot be intuitively given because it functions as the origin of any such givenness is worked out at length in 1932 in Fink's *Sixth Cartesian Meditation,* which Merleau-Ponty read in manuscript in 1942 and cites in his 1945 *Phénoménologie de la perception.*[28] Let us accordingly turn to Merleau-Ponty's appropriation—through Fink—of the notion of functioning intentionality.

---

22 "Die Aufdeckung der im Verborgenen wirkenden, sich in ihrem Resultat verhüllenden *sinnerfüllten* Bewußtseinsweisen ist die Aufgabe der intentionalen Analyse. Ihr Thema ist die *fungierende Intentionalität,* die lebendig sinnbildende, sinnleistende, sinnverwandelnde Funktion des Bewußtseins, welche zu den einfachen seelischen Einheiten der Akte sich selbst verdeckend zusammenschließt" (*ibid.*, 219).

23 *Ibid.*, 220.

24 *Ibid.*, 218, 222.

25 *Ibid.*, 222f. Fink identifies this as a question of the being of intentionality; cf. *ibid.*, 210, where he warns that the meeting of consciousness and existent cannot be forced into a subject-object scheme shaped by the model of perceiving a transcendent thing.

26 *Ibid.*, 222.

27 *Ibid.*, 205.

28 See Herman Leo Van Breda, "Maurice Merleau-Ponty et les Archives-Husserl à Louvain," *Revue de métaphysique et de morale* 67 (1962): 421f.; Maurice Merleau-Ponty, *Phénoménologie de la perception* (Paris: Gallimard, 1945), i, xv, henceforth abbreviated as *PP*.

As is well known, Merleau-Ponty's philosophy was shaped both by his acquaintance with Husserl's work,[29] especially the *Nachlass,* and by his encounter not only with Fink's writings, but with the man himself.[30] Others have documented the decisive influence of Fink's concerns on Merleau-Ponty's work in general;[31] here I am focusing only on specific references to functioning intentionality in *Phénoménologie de la perception,* references that Bruzina[32] credits with inaugurating the "prioritizing of functioning intentionality" as "one of the principal features of interest in post–World War II phenomenology." Merleau-Ponty introduces this notion in his preface (*PP*, xiii), mentioning the German term *"fungierende Intentionalität"* and translating it as *"l'intentionnalité opérante";* he contrasts it with the act-intentionality of judgments and links it with the pre-predicative unity of the pre-objective world. The relation between the two is later specified as follows:

> In Husserlian terms, we must recognize beneath the "act-intentionality" that is the thetic consciousness of the object […] an "operative" intentionality *(fungierende Intentionalität)* that makes the former possible and is what Heidegger calls transcendence.[33]

Merleau-Ponty's footnote to the term "operative intentionality" refers both to the passage from *Formal and Transcendental Logic* cited as the initial moment of the trajectory and the "landmark" passage quoted above from Fink's 1939 essay,

---

29 See Ted Toadvine, and Lester Embree, eds., *Merleau-Ponty's Reading of Husserl* (Dordrecht: Kluwer Academic Publishers, 2002).

30 Cf. Van Breda, "Maurice Merleau-Ponty et les Archives-Husserl à Louvain," 413.

31 See, e.g., Fred Kersten, "Notes from the Underground: Merleau-Ponty and Husserl's Sixth Cartesian Meditation," in *The Prism of the Self: Philosophical Essays in Honor of Maurice Natanson*, ed. Steven Galt Crowell (Dordrecht: Kluwer Academic Publishers, 1995), 43–58; Ronald Bruzina, "Eugen Fink and Maurice Merleau-Ponty: The Philosophical Lineage in Phenomenology," in *Merleau-Ponty's Reading of Husserl*, ed. Ted Toadvine and Lester Embree (Dordrecht: Kluwer Academic Publishers, 2002), 173–200; Ronald Bruzina, "Construction in Phenomenology," in *The Reach of Reflection: Issues for Phenomenology's Second Century*, ed. Steven Crowell, Lester Embree, and Samuel J. Julian (Boca Raton, FL: Center for Advanced Research in Phenomenology, www.electronpress.com, 2001), 46–71, and cf. Bruzina "Translator's Introduction," vii; xxi; lxxxiii, n. 119.

32 Bruzina, *Edmund Husserl and Eugen Fink,* 197.

33 "En langage husserlien, au-dessous de l''intentionnalité d'acte' qui est la conscience thétique d'un objet, […] il nous faut reconnaître une intentionnalité 'opérante' *(fungierende Intentionalität),* qui rend possible la première et qui est ce que Heidegger appelle transcendance" (*PP*, 478).

along with the 1928 edition of Husserl's lectures on time-consciousness.[34] No reference is provided for the remark about Heidegger.[35] However, it is at least possible that this point came up during Merleau-Ponty's conversations with Fink (mediated by Van Breda) in April 1939 in Louvain. In the course of explicating Fink's critical reconsideration of Husserl's phenomenology, Bruzina emphasizes that Fink finds it necessary to explore the kind of awareness that "has to be *already operative pre-thematically* in order for a thematic awareness to set in subsequently," illustrating this with a quotation from one of Fink's manuscripts (1930):

> The "present" [*die Gegenwart*] is always in awareness in the very performance of it [*vollzugsmäßig bewußt*], even if a-thematically. Perception breaks down into an *objective* intentionality, consciousness-of, and the *time-intentionality of making-present* [*Gegenwärtigen*]. Only on the basis of making-present can thematic intentionality be established. (See Heidegger's doctrine of "transcendence.")[36]

Moreover, Fink's 1939 reference to functioning intentionality being concealed by its results is echoed when Merleau-Ponty writes:

> Beneath act-intentionality or thetic intentionality we recover, as its condition of possibility, an operative intentionality already at work prior to any positing or any judgment, a "*logos* of the aesthetic world" … that, like all art, is known only by its results.[37]

---

34 The reference to the time lectures is to p. 430 of the original 1928 publication (*Hua* X, 76; cf. 80); the note also refers to Husserl 1929, 208 (= *Hua* XVII, 242) and to Fink 1939, 266 (= 1966, 219). Note that the English translation of Merleau-Ponty's *Phénoménologie de la perception* erroneously specifies p. 286 of the latter work; note also that *PP*, viii, n. 1, refers to Fink 1933, 331ff., but the reference should probably be to 350ff. (= 1966, 115ff., and cf. also 182ff.); finally, note that although Merleau-Ponty fails to include Husserl 1929 in his reference list, he also refers to it elsewhere (*PP*, xiv, n. 1; 255, n. 1; 453, n. 1, 3; 490, n. 2).

35 A related passage lacks not only a reference, but a specific name when Merleau-Ponty (*PP*, 141, n. 4) refers to Husserl's discovery, "beneath the intentionality of representations" (Merleau-Ponty is probably thinking of *Vorstellungen,* which in a Husserlian context should be understood as explicit "objectivating"), "of a deeper intentionality that others have called existence" ("sous l'intentionnalité des représentations, d'une intentionnalité plus profonde, que d'autres ont appelée existence").

36 Cited in Bruzina, *Edmund Husserl and Eugen Fink*, 125; this point about time is the very issue at stake in the passage Merleau-Ponty refers to in the time lectures.

37 "Nous retrouvions sous l'intentionnalité d'acte ou thétique, et comme sa condition de possibilité, une intentionnalité opérante, déjà à l'œuvre avant toute thèse ou tout jugement, un 'Logos du monde esthétique' […] qui, comme tout art, ne se connaît que dans ses résultats" (*PP*, 490f.). In a footnote crediting Husserl 1929, 257 (= *Hua* XVII,

In addition, however, another reference to such concealment indicates a further possible inspiration for Merleau-Ponty's thinking on these matters:

> It is not easy to lay bare pure motor intentionality: it is hidden behind the objective world it contributes to constituting.[38]

The specific mention of "motor intentionality" in this citation recalls certain passages in the parts of Husserl's *Crisis* to which Merleau-Ponty had access in manuscript form during his spring 1939 visit to Louvain.[39] Here Husserl refers in §28 to "kinaesthetically functioning Corporeality" ("kinästhetisch fungierenden Leiblichkeit")—i.e., "the egoically active functioning of the Body or Bodily organs" ("das ich-tätige Fungieren der Leibes bzw. der Leibesorgane")—implicated in bringing things to perceptual appearance even if such activities and habitualities of the functioning I remain unthematic (*Hua* VI, 109; cf. 111). In any case, the notion of functioning Corporeality enters the public tradition with the publication of *Phénoménologie de la perception,* nine years before the publication of the full *Crisis* text in *Husserliana* VI in 1954. And Merleau-Ponty's subsequent thought not only continued to make explicit reference to "the domain of the *fungierende,*"[40] but radicalized the notion of an operatively functioning Body with his appeal to the innate "anonymity" he terms "flesh."[41]

---

297) for the phrase in quotation marks, Merleau-Ponty is careful to remind us that this is a matter of the transcendental aesthetic, but cf. *L'Œil et l'Esprit* (Paris: Gallimard, 1964), 29f., on the play of shadows in Rembrandt's *Nightwatch* that makes the painted object visible while itself remaining concealed. The theme of that which functions operatively but is concealed by its results is also found in Maurice Merleau-Ponty, *Notes de cours 1959–1961*, ed. Stéphanie Ménasé (Paris: Gallimard, 1996), 76: "Forgottenness of the operation of *Stiftung* [primal institution], masked by its own results" ("Oubli de l'opération de *Stiftung* masquée par ses propres résultats").

38 "Il n'est pas facile de mettre à nu l'intentionnalité motrice pure: elle se cache derrière le monde objectif qu'elle contribue à constituer" (*PP*, 161, n. 1); on motor intentionality, see also 128; 160; 222, n. 2; 290; 366; 444.

39 Only Parts I and II of this work were published in 1936; as Van Breda points out ("Maurice Merleau-Ponty et les Archives-Husserl à Louvain," 415), Merleau-Ponty cites the manuscripts he read as Parts II and III (see *PP*, 523), but they were actually Parts III A and III B.

40 Merleau-Ponty, *Notes de cours 1959–1961*, 381 (and cf. 379); see also "Husserl aux limites de la phénoménologie. 1960," ed. Franck Robert, in Merleau-Ponty, *Notes de cours sur L'Origine de la géométrie*, suivi de *Recherches sur la phénoménologie de Merleau-Ponty,* sous la direction de Renaud Barbaras (Paris: Presses Universitaires de France, 1998), 77 (and cf. 82; 85, n. 3), as well as *Le visible et l'invisible*, ed. Claude Lefort (Paris: Gallimard, 1964), e.g., 158f., 292, 297f. (and cf. 299).

41 See, e.g., Merleau-Ponty, *Le visible et l'invisible*, esp. 183; cf. Bruzina, "Construction in Phenomenology," 66ff.

It is not possible to discuss these developments here. However, before turning to the next "landmark" on the trajectory I am tracing, it is appropriate to identify a specific circumstance characterizing both Fink's and Merleau-Ponty's contributions to the efficient history of the "functioning." This involves what might be termed "insider" access to Husserl's *Nachlass* in general (and to his research manuscripts—*Forschungsmanuskripten*—in particular), in the case of Fink through living conversation[42] as well as through work on the manuscripts, and in the case of Merleau-Ponty through his 1939 visit to Louvain.[43] In contrast, the works by Brand and Held to be considered below are based on what might be termed "archival" access to the *Nachlass* materials in general (and especially to the research manuscripts), with each author visiting a branch of the Husserl Archives during the course of his dissertation research[44] and thanking Van Breda as Director of the Archives.[45] Each of these authors can be credited with shaping the tradition of Husserl-interpretation by bringing his research theme—world-experiencing life for Brand, living present for Held—to the attention of other scholars through their publications. But they are part of the efficient history of the notion of functioning subjectivity as well.

If we now turn to Brand's 1955 *Welt, Ich und Zeit,* two points about its status as a "landmark" in the trajectory we are exploring immediately stand out: it includes a section (§6) explicitly introducing functioning intentionality as a concept that will be used throughout, and material from this section was sub-

---

42 Cf. Bruzina, *Edmund Husserl and Eugen Fink*, 7.

43 In a letter to Merleau-Ponty before his visit, Van Breda ("Maurice Merleau-Ponty et les Archives-Husserl à Louvain," 413) notes that it might not be possible for Merleau-Ponty to obtain permission to cite the unpublished manuscripts he could see there, since the rules for such use had not yet been determined; it is significant that Merleau-Ponty's reference list (*PP*, 523) thanks Monsignor Noël and the Institut Supérieur de Philosophie de Louvain, rather than the Husserl Archives, for permission to cite the three unpublished texts that he lists, even though the Archives had already been established in principle in 1938—see Herman Leo Van Breda, "Le sauvetage de l'héritage husserlien et la fondation des Archives-Husserl" in *Husserl et le pensée moderne*, ed. Herman Leo Van Breda and Jacques Taminiaux (LaHaye: Martinus Nijhoff, 1959), 34.

44 According to Gabel, Brand's dissertation was submitted in 1950—Gernot U. Gabel, *Husserl. Ein Verzeichnis der Hochschulschriften aus westeuropäischen und nordamerikanischen Ländern 1912–1990* (Hürth: Edition Gemini, 1995), 16. Klaus Held (Lebendige Gegenwart, xiii) mentions 1962 as the date of his own dissertation, although Gabel (*Husserl*, 20) gives 1963. The discrepancy might reflect, for instance, a difference between date of completion and date of formal submission.

45 See Brand, *Welt, Ich und Zeit*, xvi; Held, *Lebendige Gegenwart*, xiv, where the now customary formula acknowledging permission to cite unpublished texts is used.

sequently included in a widely available edited collection[46] in English translation, fleshing out the little that was already available in English on this theme in the 1962 English translation of Merleau-Ponty's *Phénoménologie de la perception* (the English translation of the *Crisis* did not appear until 1970). The context in which the concept in question is initially introduced has to do with the dynamic, self-transcending character of intentionality; then Brand goes on to say, "It doesn't simply *exist*—rather, it *functions;* that is why Husserl calls it 'functioning intentionality,'" with a footnote to Fink's 1939 article.[47] In Part I of the work, Brand accordingly carries out an intentional analysis that takes the experienced horizonality of the world as a leading clue *(Leitfaden),* inquiring back into the experiencing, self-transcending consciousness that is characterized not only by intentionality in general in that it is conscious-of something other than itself, but also by horizon-intentionality in particular in that it always means or projects more than what is currently actually given. Thus for Brand,[48] the "genuine transcendental problem" is the question of the functioning intentionality that performs such achievements. His initial account emphasizes functioning intentionality as functioning anonymously while we are straightforwardly directed toward entities as objects standing over-against the anonymous I who is occupied with these objects.[49] It is only when transcendental phenomenology inquires back into what functioning intentionality has always already anonymously achieved—the ready-made unities of natural attitude experience[50]—that the ultimately functioning/achieving subjectivity is made visible and comprehensible, freed from its anonymity and recognized for what it is.[51] This subjectivity does not somehow first arise via the phenomenological reduction, but is already continuously functioning in the natural attitude[52] and has a prior understanding of itself that is presumed in any subsequent phenomenological explication of it,[53] even if such functioning only comes to awareness in a peculiar way that is yet to

46 Joseph J. Kockelmans, ed., *Phenomenology: The Philosophy of Edmund Husserl and Its Interpretation* (Garden City, NY: Doubleday, Anchor Books, 1967).

47 "Sie *ist* nicht schlechthin, sondern sie *fungiert.* Deswegen nennt Husserl sie 'fungierende Intentionalität'" (Brand, *Welt, Ich und Zeit*, 23), referring to Fink, "Das Problem der Phänomenologie Edmund Husserls," 266, the same page to which Merleau-Ponty (*PP*, 478, n. 1) refers, i.e., Fink, *Studien zur Phänomenologie 1930–1939*, 219 (see n. 22 above).

48 Brand, *Welt, Ich und Zeit*, 45.

49 Cf., e.g., Brand, *Welt, Ich und Zeit*, 24ff.

50 *Ibid.*, 38.

51 *Ibid.*, 42ff.; cf. 35f.

52 *Ibid.*, 46.

53 *Ibid.*, 51.

be specified: "It is indeed conscious and makes [us] conscious [of something], but we do not already experience it, we are not already [explicitly] aware of it,"[54] i.e., we have a certain awareness of ourselves as living in our acts, but without normally taking functioning subjectivity as a thematic object.[55]

Then in Part II of his work Brand explicitly turns to the question of making this "implicit" mode of constitutively functioning subjectivity thematic.[56] The key passages are to be found in the discussions of reflection in §§12–14, where he distinguishes two senses of anonymous functioning: the self-transcending, sense-bestowing, world-projecting intentionality addressed in Part I, and a "more primitive" anonymous functioning in the sense of the sheer "being active" of an I that functions as a living pole whose self-presence remains anonymous while everything stands over-against it as an object.[57] Even when it thematizes its own functioning in reflection, all it finds is the just-past functioning while the current reflecting remains anonymous, and so on.[58] But this means that the ultimate transcendental functioning never catches up with itself, since all that can be thematized is a past, hence temporally individuated, hence mundane I,[59] while the most primal mode of being of world-experiencing life is its functioning in the "functioning-anonymous" present.[60] We may summarize Brand's contribution to the development of the notion of functioning subjectivity by saying that functioning intentionality makes reflection on itself possible insofar as we have a kind of pre-thematic acquaintance with this functioning subjectivity, but it itself is never definitively objectified *in* its functioning—all we have is a coincidence *(Deckung)* between the past reflected-upon I and the present reflecting I. Thus the problem becomes the essential undisclosability of the living functioning through which everything comes to presence—a problem that continues to haunt the efficient history of the notion of functioning subjectivity. Before considering Held's contribution to this tradition, however, it is appropriate to turn to the further influence of Eugen Fink through his 1957 essay on "operative" concepts in Husserl, originally delivered as a lecture at the "III. Colloque International de Phénoménologie" in Royaumont.

54 "Sie ist zwar bewusst und macht bewusst, aber ist darum nicht schon erfahren und gewusst" (*ibid.*, 28).

55 *Ibid.*, 58.

56 See *ibid.*, 24, citing Ms. [cf. text p.25 and p.25, n.77] A VII 9, and cf. *Hua* XXXIX, 782, on the contents of this group of manuscripts, which also address the question of the distinction between such functioning subjectivity and real/personal mundane subjectivity.

57 *Ibid.*, 62ff.

58 Cf. *ibid.*, 65f.

59 Cf. *ibid.*, 46, 51, 97f., 135, 140.

60 *Ibid.*, 84.

The first question concerns Fink's title: why does he use the German word *"operative"* rather than *"fungierende"?* It could have to do with the linguistic context of the venue and with the fact that Merleau-Ponty had used *"opérante"* as a translation for *"fungierende"* in *Phénoménologie de la perception,*[61] or it may be that Fink deliberately chose a different term in order to separate his notion from Husserl's. A second issue concerns the relation(s) between operative experiencing *(Erleben)* and operative concepts.[62] Neither of these points can be decided here, but they need not be resolved in this context, where the task is simply to consider the contribution this essay makes to the efficient history of the notion of the "functioning." Fink's key distinction is between the *"fundamental thematic concepts"* at the heart of a philosopher's work—e.g., Nietzsche's "will to power" or Husserl's "transcendental subjectivity"—and the operative concepts "through which" the philosopher thinks, concepts with which the thinker "operates" without thematizing these concepts in their own right: "[The philosophers'] conceptual understanding moves in a *conceptual field,* a *conceptual medium,* that they themselves are not at all capable of bringing into view."[63] Fink goes on to develop the visual metaphor, referring to such concepts as the "shadow" of a philosophy.[64] But what is decisive is that for Fink, the phi-

---

61 Note also that the subentry under "Intentionality" in Spileers' "Index of Keywords" is "operative intentionality," while the title of the sole work listed refers to "Fungerende intentionaliteit," indicating a tendency toward the terms becoming interchangeable—see Steven Spileers, comp. *Edmund Husserl Bibliography. Husserliana Dokumente IV* (Dordrecht: Kluwer Academic Publishers, 1999), 401; Antoine Kortooms, "Fungerende intentionaliteit bij Husserl," in *Gehelen en fragmenten. De vele gezichten van de filosofie*, ed. Bart Raeymaekers (Leuven: Universitaire pers Leuven, 1993), 124–129.

62 Cf. Fink, "Operative Begriffe in Husserls Phänomenologie," 329f.; J. N. Mohanty, "Understanding Husserl's Transcendental Phenomenology: An Introductory Essay," in *Apriori and World: European Contributions to Husserlian Phenomenology*, ed. and trans. William McKenna, Robert M. Harlan, and Laurence E. Winters (The Hague: Martinus Nijhoff, 1981), 5f.

63 "Ihr begriffliches Verstehen bewegt sich in *einem Begriffsfeld,* in einem *Begriffsmedium,* das sie selber gar nicht in den Blick zu nehmen vermögen" (Fink, "Operative Begriffe in Husserls Phänomenologie," 325).

64 One is immediately reminded of the title of Merleau-Ponty's 1959 essay ("Le philosophe et son ombre"), although Merleau-Ponty refers only to Fink's "L'analyse intentionnelle et le problème de la pensée spéculative," in *Problèmes actuels de la phénoménologie*, ed. Herman Leo Van Breda (paris: Desclée de Brouwer, 1952), 54–87, and does not cite Fink's "Operative Begriffe in Husserls Phänomenologie," referring instead (as he also does in "Husserl aux limiter de la phénoménologie. 1960," 14) to Heidegger's notion of the *Ungedachte;* the resonance between Fink and Heidegger here can be gauged by comparing Fink, "Operative Begriffe in Husserls Phänomenologie," 336—"The more original the force of [what a philosophy brings to light], the deeper are

losophers are *unable to see* the operative concept that allows them to philosophize at all:

> It is not "thematic" because it is that through which we are related to the theme. It is what is not seen because it is the *medium of seeing*.[65]

Thus everything hinges not merely on a distinction between the operative and the thematic, but on a tension between them, and Fink identifies this very tension as a theme of Husserlian thinking itself,[66] referring to Husserl's attempts to conceptualize that which is lived through without being grasped in the manner of a thematic object.[67] Fink goes on to specify which phenomenological concepts he finds Husserl operatively using rather than thematically clarifying[68] during the course of this attempt to thematize the operative presuppositions at work in the natural attitude. But perhaps the most telling way in which this article shapes and sustains the trajectory we are tracing is through the notion of an *essential unthematizability* that we cannot overcome any more than we can leap over our own shadow,[69] concluding that this is the fate of any finite philosophizing.[70] And in Held's well-known work on the living present, the notion of that

---

the shadows in [its] fundamental ideas as well" ("Je ursprünglicher die Kraft ist, die eine Lichtung wagt, desto tiefer sind auch die Schatten in den Grundgedanken") with the passage from Martin Heidegger, *Der Satz vom Grund* (Pfullingen: Neske, 1957), 123f., that Merleau-Ponty cites: "The greater a thinker's thought is, […] the richer is that which is unthought in this thought" ("Je größer das Denkwerk eines Denkers ist, […] um so reicher ist das in diesem Denkwerk Ungedachte […]"). However, Merleau-Ponty's explication of the "shadow" is oriented toward bringing out further elements suggested by, but not yet worked out in, a given thinker's work; moreover, he compares them to horizons and levels as that through which the appearing appears, i.e., he is referring to moments on the side of the perceptual world, not to operative performances of subjectivity. His tendency to think a notion derived from the subjective dimension in terms of "external being" also surfaces in "Husserl aux limites de la phénoménologie. 1960," 85, n. 3.

65 "Es ist nicht 'im Thema,' weil wir durch es hindurch uns auf das Thema beziehen. Es ist das Nichtgesehene, weil es das *Medium des Sehens* ist" (Fink, "Operative Begriffe in Husserls Phänomenologie," 327; cf. *Hua* XVII, 38).

66 Fink, "Operative Begriffe in Husserls Phänomenologie," 327.

67 *Ibid.*, 328f.

68 *Ibid.*, 330ff. Note that Fink (*ibid.*, 333, 334, 335) refers to Husserl's transformation of mundane concepts into transcendental ones as "speculative."

69 *Ibid.*, 325. Fink, "Die Spätphilosophie Husserls in der Freiburger Zeit," in *Edmund Husserl 1859–1959*, ed. Herman Leo Van Breda and Jacques Taminiaux (La Haye: Martinus Nijhoff, 1959), 114, uses a different metaphor in referring to the difficulty of catching the operative "red-handed" ("'auf der frischen Tat zu ertappen'"), as one would catch a thief in the act of stealing.

70 Fink, "Operative Begriffe in Husserls Phänomenologie," 336.

which not only functions operatively but cannot *in principle* be reflectively thematized is deepened still further; in fact, one gains the impression that while Brand uses the word "functioning" as an adjective, Held transforms the term into the proper name *(Eigenname)* of a problem.

The first point to make, however, is that by the time Held's work appears in 1966, the problematic of functioning subjectivity has already begun to coalesce, as it were, in the literature. For example, Held's reference list includes all of the "landmarks" already discussed above as moments in the efficient history of this notion.[71] But in addition, the published version of *Lebendige Gegenwart* incorporates some critical remarks on another dissertation that had been published only after Held had completed the main text of his own work—remarks where what is at stake is precisely the issue of whether functioning subjectivity can, *as functioning,* become an object.[72] It is also important to note the publication of Husserl's 1923/24 lectures *(Erste Philosophie)* in two parts in 1956 and 1959, and especially the appearance of the second part, *Theorie der phänomenologischen Reduktion* (*Hua* VIII).[73] The latter volume is certainly relevant to Held's project, even if other terminology is often used instead of "functioning."[74] However, it is not merely that this volume is available for him to cite; instead, as he tells us,[75] his work was significantly shaped by his teacher Landgrebe's 1962

71 See Held, *Lebendige Gegenwart*, xii, for some critical remarks concerning Brand, *Welt, Ich und Zeit.*

72 See Held, *Lebendige Gegenwart*, xiii, for his basic disagreement with Thomas Seebohm, *Die Bedingungen der Möglichkeit der Transzendental-philosophie* (Bonn: Bouvier, 1962), the published version of the latter's 1960 dissertation; for specific points, cf., e.g., ibid., 66f., 138, 161; Held, *Lebendige Gegenwart*, 76ff., 125.

73 Boehm's editorial introduction to this volume (*Hua* VIII, xi) relates an oral communication from Ingarden indicating that although Landgrebe had typed the main text of the lectures by 1924, Husserl apparently did not share this material with friends and students (as had been the case with other texts); thus they did not immediately influence those with "insider" access in the same way that, e.g., *Ideen II* did.

74 The term *"fungierende"* does appear in the supplementary materials (see *Hua* VIII, Beilage XXI, esp. 439ff., and cf. 478), but Husserl also contrasts the "latent" I with the "patent" I (see, e.g., *Hua* VIII, 90, 408f. as well as Beilage XXI) and speaks of overcoming what he at first calls the "self-forgetfulness" of the I (but cf. *Hua* VIII, 90) through the "splitting" *(Spaltung)* of the I accomplished in reflection; on the latter, see, e.g., Jan M. Broekman, *Phänomenologie und Egologie. Faktisches und transzendentales ego bei Edmund Husserl* (Den Haag: Martinus Nijhoff, 1963), 122–133, and cf. Gisbert Hoffmann, "Die Zweideutigkeit der Reflexion als Wahrnehmung von Anonymität," *Husserl Studies* 14 (1997): esp. 108, 113ff.; Luft, *"Phänomenologie der Phänomenologie,"* section 2.2 (and cf. 3.4).

75 Held, *Lebendige Gegenwart*, xiii.

essay "Husserls Abschied vom Cartesianismus," a study of this very volume that attempts to clarify some of its operative concepts, emphasizing the same issues that Held himself takes up in his treatment of the functioning I. Moreover, Held[76] also thanks Landgrebe for a 1960/61 university colloquium interpreting Ms. C 2 I, which includes material on functioning subjectivity;[77] such formal attention to this manuscript (and the matters it refers to) clearly signals that problems connected with the interpretation of the notion of functioning subjectivity are emerging as a coherent theme within the generative phenomenological tradition, and the efficient history of this notion is an increasingly public history as *Nachlass* materials are more widely disseminated and discussed. Finally, it is significant that when Held introduces first the idea of the "functioning" transcendental life[78] and then the need to free this transcendental life from its "anonymity,"[79] both passages identify the terms in question as Husserlian locutions by using the phrase "as Husserl says," without any supporting page references: the notion of anonymously functioning subjectivity is presented as a recurring working notion of Husserl's thinking, rather than as a concept or description stemming from a particular passage in a particular context. Of course, Held's book provides many exact quotations from Husserl's research manuscripts, making them available to an even wider audience. But he also makes further contributions to the trajectory leading to what will eventually become something like a "received interpretation" of the notion of functioning subjectivity by carrying forward Brand's work in several main ways.[80] First, Held fine-tunes and focuses the idea that inquiry back into anonymously functioning subjectivity takes two forms: we can bring to light the normally concealed constitutive achievements through which we experience a transcendent world, or we can penetrate more deeply into the self-temporalization that makes such achievements possible at all.[81] Furthermore, he systematically sets forth an interpretation of reflection in

76 *Ibid.*, xiv.

77 See *Hua Mat* VIII, Texts Nr. 1–4, esp. 1 and 3, along with *Hua* XXXIV, Text Nr. 20. Other participants in the colloquium included Karl-Heinz Volkmann-Schluck and Walter Biemel (Held, *Lebendige Gegenwart*, xiv), and in section V of his important 1963 essay on the phenomenological movement, Gadamer thanks Landgrebe and Volkmann-Schluck for the fact that he has a copy of Ms. C 2 I to refer to as he comments on Fink's Husserl-interpretation.

78 Held, *Lebendige Gegenwart*, 4.

79 *Ibid.*, 7.

80 Those familiar with Held's work will undoubtedly agree that what is said here does not do justice to his detailed analyses or his careful attempts to solve the riddles raised by the points I highlight; as mentioned, however, my task is to document a historical trajectory rather than to provide detailed interpretations of any of its "landmark" texts.

81 See, e.g., Held, *Lebendige Gegenwart*, 89f., 101ff.; cf. 120.

terms of an irrevocable "distance" between the temporalizing I on the one hand and the temporalized I that is grasped as an object over-against it on the other, so that the subsequent reflection never catches up with the original functioning.[82] Finally, he confirms the uncancellable experiential inaccessibility of the functioning that makes experience and its achievements possible in the first place[83]—and in the process, he challenges Husserl's fundamental method:[84] it is necessary to go beyond the realm of what can be demonstrated through the intuitive grasp, in reflection, of something originally given as "itself there."[85] In addition, however, Held adds a particular nuance to the notion of the limits of reflection by identifying the visual paradigm at its heart. The structure of reflective "seeing" involves directing the intentional regard *(Blickstrahl)* from the functioning I (the I of the original functioning present) as its origin toward the object-I as its terminus. But a distance between seeing and seen is always already presupposed in the very attempt to see; the self that is grasped is no longer the self carrying out the grasping:[86]

> In other words, what reflection ultimately winds up with is now the insight that I cannot catch sight of and grasp myself because I myself am the source of my own functioning—I myself am the grasping regard.[87]

And once again we find an author explaining that the I cannot leap over its own shadow in order to experience the continually and uncancellably anonymous "I function" in the act.[88]

At this point, it is appropriate to turn to yet another type of "landmark" in the history of functioning subjectivity, one appearing in two forms in 1971 and 1972. In his 1962 dissertation (published in 1963), Jan Broekman[89] is critical of the phenomenological literature for failing to address Husserl's theory of intentionality in sufficient detail; he points out, for example, that a 1959 introduction to Husserl's phenomenology does not even discuss functioning intentionality,

---

82 *Ibid.*, 9, 40, 63, 74, 81, 131.

83 See, e.g., *ibid.*, 95, 103, 112f., 115, 118f., 123, 133, 134.

84 *Ibid.*, 104, 117, 137.

85 *Ibid.*, 145; cf. x–xi, 150.

86 *Ibid.*, 119f.

87 "Mit anderen Worten: das nunmehr Letzte der Reflexion ist die Einsicht: Ich kann mich selbst nicht erblicken und erfassen, weil ich selbst das Entquellen meines Fungierens,—weil ich der erfassende Blick selbst bin" (*ibid.*, 120; cf. Luft, *"Phänomenologie der Phänomenologie,"* 107ff.).

88 Held, *Lebendige Gegenwart*, 133. Held does not refer to the use of this metaphor in Fink, "Operative Begriffe in Husserls Phänomenologie," 325, but the latter work is included in his reference list.

89 Broekman, *Phänomenologie und Egologie*, 47ff.

although significant mention of it is to be found in Husserl's writings.[90] By 1972, however, the growing literature of the phenomenological tradition includes a now classic monograph—Mohanty's *The Concept of Intentionality* (the full version that is summarized in "Husserl's Concept of Intentionality," published in 1971)—that supplies what Broekman had called for, including a brief but cogent discussion of "'operative intentionality'"[91] as a title uniting such notions as passive synthesis, horizon intentionality, genetic constitution, pre-predicative intentionality, and anonymous intentionality.[92] Mohanty[93] refers to key passages from most of the "landmark" works already discussed (*Formal and Transcendental Logic,* Fink's 1939 article, Merleau-Ponty's *Phénoménologie de la perception,* Brand's 1955 book), concluding as follows:

> In fine, we could say that the notion of operative intentionality gathers together the various strands of thought which emerge in Husserl's writings at various stages. We have at first the thesis that intentionality constitutes objects, and also the further thesis that not all constitution is active and conscious. These two lead to the position that there is a passive synthesis which, itself hidden and anonymous, constitutes transcendence. Try to uncover this anonymous operation, and you are led from one layer to another till you reach the last ground of all synthesis: this is the world-experiencing-life, the source of all meanings and constitution [...].
>
> Thus, the expression 'operative intentionality' is used to designate two very different things: first the intentionalities other than acts, which are passive, anonymous

90 *Ibid.*, 48f. The work he is critical of in this regard is Wilhelm Szilasi, *Einführung in die Phänomenologie Edmund Husserls* (Tübingen: Max Niemeyer, 1959). Broekman himself relies on published works for his understanding of functioning subjectivity (see, e.g., *Phänomenologie und Egologie*, 210, citing *Hua* VI, 102); no *Nachlass* materials are included in his reference list, and what citations there are (e.g., from K manuscripts) are taken from Brand, *Welt, Ich und Zeit.* This is one of many possible examples of the dissemination of certain research manuscripts through citations of these materials in the secondary literature, making such texts available for interpretation by other scholars writing dissertations on specialized themes; cf. Saulius Geniusas, "The Post-War Reception of *Ideen I* and Reflection," in *Husserl's Ideen*, ed. Lester Embree and Thomas Nenon (Dordrecht: Springer, 2013), 400ff., on reception patterns in general.

91 Mohanty's choice of the term "operative" is probably influenced by Merleau-Ponty, and in addition allows him to make a clear distinction between theories of consciousness emphasizing intentionality and the claims of a "functionalist" psychology. See Mohanty, *The Concept of Intentionality*, 184f.

92 *Ibid.*, 121; see 116–123 for the relevant discussions.

93 Mohanty, "Husserl's Concept of Intentionality," 114f.; *The Concept of Intentionality*, 122.

> and synthetic; but also the supposed last source of all meaning and all constitution: the world-experiencing-life.[94]

It is striking that the distinction between the two meanings of operative intentionality corresponds to the points in Brand and Held concerning two different forms of anonymity, yet Mohanty's exposition of Husserl's work does not emphasize any difficulty with gaining phenomenological access to the second, ultimately functioning intentionality.[95] Instead, such issues as a pre-reflective Bodily intentionality that is open to the world without being aware of itself are linked with the existentialist interpretation of intentionality, and above all with Merleau-Ponty's appropriation of the notion of operative intentionality.[96] But in any case, Mohanty's 1972 presentation signals that the notion of functioning subjectivity, here represented by the concept of operative intentionality, has become an "official" working notion in the phenomenological literature, as we can also see, for instance, from a work published the next year. Hans-Ulrich Hoche's attempt to work out a purely noematic phenomenology in his *Handlung, Bewusstsein und Leib* (1973) simply makes use of the notion of a functioning Body, introduced without comment and employed without ever referring to any of the "landmark" moments in the trajectory we have been following: it has become part of a received tradition and serves as a ready-made concept that the author can draw upon in his critical discussions of Sartre and Husserl.[97] This concludes my "landmark-by-landmark" account of the trajectory leading to this concept, since the same year—1973—may be seen as the beginning of a new interpretive tradition based on the increasing public availability of *Nachlass* materials in the *Husserliana* series. But there is one more moment to mention in the

---

94 Mohanty, *The Concept of Intentionality*, 123.

95 This may be because Mohanty has a further distinction at his disposal: namely, the distinction between "intentionality" as directedness toward an object and "reflexivity" as a style of subjective self-awareness irreducible to intentionality, providing an alternative to the notion of "reflection" as a process where consciousness turns its intentional regard back upon itself and makes itself an "object" (albeit in the mode "just past"); see Mohanty, "Husserl's Concept of Intentionality," 126ff.; *The Concept of Intentionality*, Part Three, Ch. 1.

96 Mohanty *The Concept of Intentionality*, 137ff.

97 See, e.g., Hans-Ulrich Hoche, *Handlung, Bewußtsein und Leib. Vorstudien zu einer rein noematischen Phänomenologie* (Freiburg: Karl Alber, 1973), 27, 33; §§26–28; 324ff. Ullrich Melle, *Das Wahrnehmungsproblem und seine Verwandlung in phänomenologischer Einstellung. Untersuchungen zu den phänomenologischen Wahrnehmungstheorien von Husserl, Gurwitsch und Merleau-Ponty* (Den Haag: Martinus Nijhoff, 1982), §14—"Vom noematischen zum noetischen Begriff des leiblichen Bewusstseins"—critically addresses Hoche's approach to the functioning Body.

arc of the efficient history through which a received interpretation of functioning subjectivity has continued to make its mark in the literature.

If we review the tradition documented so far, we can see that there is a shift from a living intentionality that "may" function unthematically to one that "must" function unthematically—a shift from a functioning subjectivity that is concealed in the natural attitude (which knows only its results) but revealed in phenomenological analysis to a functioning subjectivity that cannot in principle be grasped, in its functioning, by the reflective regard. This shift is accomplished in part by the distinction between two forms of the anonymity pertaining to functioning subjectivity. In a sense, we may see Fink as the architect laying out the main lines of this shift, while Brand and Held both furnish the detailed scholarship on the basis of which the philosophical implications of the shift become clear and bring these matters into the realm of public discourse. But on my reading, it is Merleau-Ponty's widely disseminated appropriation of the concept[98] that has exerted the most influence on the tendency to take operative intentionality in general—and the operative Body in particular—as a pre-reflective power or performance functioning in a way that remains fundamentally unthematic from beginning to end, and indeed, in a way that "precludes" any such thematization.[99] One example will suffice not only to illustrate this tendency, but to highlight one of its important implications. In several articles published in 1986, Shaun Gallagher draws a clear-cut distinction between a body "image" as something that can indeed come to conscious awareness and a body "schema" through which the body is not at all an intentional object, but is "absently available" in a way that is totally tacit, anonymous, pre-reflective, pre-personal, preconscious or non-conscious, pre-subjective, and pre-objective or non-

98 Note that while some phenomenologists do mention the role Fink played in Merleau-Ponty's appropriation of the concept—see, e.g., Bernhard Waldenfels, *Phänomenologie in Frankreich* (Frankfurt am Main: Suhrkamp, 1987), 42; *Einführung in die Phänomenologie* (München: Wilhelm Fink, 1992), 41—and others refer both to Husserl and to Fink—see, e.g., Natalie Depraz, "What about the *praxis* of Reduction? Between Husserl and Merleau-Ponty," in *Merleau-Ponty's Reading of Husserl*, ed. Ted Toadvine and Lester Embree (Dordrecht: Kluwer Academic Publishers, 2002) 119—there is a general tendency to associate the concept with Merleau-Ponty.

99 Depraz, "What about the *praxis* of Reduction? Between Husserl and Merleau-Ponty," 119. Depraz goes on to say, "Like Levinas later on, Merleau-Ponty claims very early that thematization destroys the richness of our operative/operant experience, which is the sole genuine phenomenological one," then (120) links this stance with Merleau-Ponty's rejection of a philosophy of reflection and proposes (120ff.) an approach to the phenomenological reduction that has been freed from the "fruitless" assumption that "thematization" is "opposed to operativity" (121).

objectifying;[100] he accordingly contends that "there is no way to describe the body in its anonymous performances in purely transcendental terms. The lived body in its absently available performances must, of necessity, escape Husserl's transcendental reflections."[101] And the implication I see here is that Husserlian phenomenology is therefore assumed to be essentially in need of supplementation, not merely through the speculative, constructive moments that are already found in both Fink and Merleau-Ponty,[102] but in a further direction as well: the operative is now merged with the physiological,[103] opening the way for double-sided inquiries in which phenomenological description on the one hand and naturalistic cognitive science, neurophysiology, etc., on the other hand interrelate in a variety of ways still open to debate.[104] It would certainly be possible to find

---

100 See Shaun Gallagher, "Hyletic experience and the lived body," *Husserl Studies* 3 (1986): 154, and cf., e.g., 146ff.; "Lived Body and Environment," *Research in Phenomenology* 16 (1986): 144ff., 153, 158; "Body Image and Body Schema: A Conceptual Clarification," *Journal of Mind and Behavior* 7 (1986): 548, 551.

101 Gallagher, "Hyletic experience and the lived body," 154.

102 See Fink, "L'analyse intentionnelle et le problème de la pensée spéculative," *VI. Cartesianische Meditation; Teil 1*; Bruzina, "Construction in Phenomenology"; Luft, *"Phänomenologie der Phänomenologie,"* section 3.5.2, esp. 195.

103 See Dan Zahavi, *Self-Awareness and Alterity: A Phenomenological Investigation* (Evanston, IL: Northwestern University Press, 1999), 240, n. 24, for critical remarks concerning Gallagher's including physiological processes in the notion of the body schema, although it can also be argued that not all operative performances of the Body are reducible to sheerly physiological "functions"—cf. Shaun Gallagher, "Body schema and intentionality," in *The Body and the* Self, ed. J. Bermudez, A. Marcel, and N. Eilan (Cambridge, MA: MIT Press, 1995), 225–244.

104 See, e.g., the sources mentioned in Tom Froese and Shaun Gallagher, "Phenomenology and Artificial Life: Toward a Technological Supplementation of Phenomenological Methodology," *Husserl Studies* 26 (2010): 86; cf., e.g., Nader N. Chokr, "Mind, Consciousness, and Cognition: Phenomenology vs. cognitive science," *Husserl Studies* 9 (1992): 179–197; Francisco J. Varela, "The Naturalization of Phenomenology as the Transcendence of Nature: Searching for generative mutual constraints," *Alter* 5 (1997): 355–381; Shaun Gallagher and Francisco Varela, "Redrawing the Map and Resetting the Time: Phenomenology and the Cognitive Sciences," in *The Reach of Reflection: Issues for Phenomenology's Second Century*, ed. Steven Crowell, Lester Embree, and Samuel J. Julian (Boca Raton, FL: Center for Advanced Research in Phenomenology, www.electronpress.com, 2001), 17–45, Jean-Luc Petit, "On the relation between recent neurobiological data on perception (and action) and the Husserlian theory of constitution," *Phenomenology and the Cognitive Sciences* 2 (2003): 281–298; Ronald Bruzina, "Phenomenology and Cognitive Science: Moving Beyond the Paradigms. Review Essay: Jean Petitot, Francisco J. Varela, Bernard Pachoud, Jean-Michel Roy, eds. *Naturalizing Phenomenology: Issues in Contemporary Phenomenology and Cognitive Science," Husserl Studies* 20 (2004): 43–88; Steven Horst, "Phenomenology and psychophysics," *Phe-*

further examples in which the received tradition of interpreting Husserl's notion of functioning subjectivity has undoubtedly been of influence. But how does it compare with Husserl's own way(s) of working with and through this notion?

In my view, this question cannot adequately be answered solely on the basis of works Husserl published during his lifetime, but requires turning to his *Nachlass,* and above all to his research manuscripts. However, more is at stake than merely locating relevant passages (and a preliminary survey reveals that there are many such passages). First of all, it is necessary to address issues of methodological hermeneutics that may arise when one is interpreting such material.[105] But in addition, it is crucial to develop the appropriate modes of access to the experiential evidence proper to the matters in question; in the present case, this means cultivating a mode of "reflexivity" that is irreducible to intentionality[106] and thus does not fall prey to the same difficulties as the "reflection" in which the intentional object can only be a past moment of a functioning intentional life whose current functioning must inevitably remain anonymous. Only when these two methodological elements are in place will it be possible to offer an account of Husserl's own notion of functioning subjectivity and to compare it with the received tradition that has dominated phenomenological discourse for so long.

---

nomenology and the Cognitive Sciences 4 (2005): 1–21; Shaun Gallagher, "Neurophilosophy and Neurophenomenology," in *Phenomenology 2005,* Vol. 5: *Selected Essays from North America*, ed. Lester Embree and Tom Nenon (Bucharest: Zeta Books, 2007), 293–316; Vincenzo Costa, "Die Erfahrung des Anderen. Phänomenologie, Behaviorismus und Spiegelneuronen," *Husserl Studies* 24 (2008): 231–241; Dan Zahavi, "Review Article: Thompson, Evan. *Mind in Life: Biology, Phenomenology, and the Sciences of Mind,*" *Husserl Studies* 25 (2009): 159–168; Shaun Gallagher, "Cognitive Science," in *The Routledge Companion to Phenomenology*, ed. Sebastian Luft and Søren Overgaard (London: Routledge, 2012), 574–585, and see also Natalie Depraz, "L'empiricité transcendantale de la phénoménologie," ch. 5 in her *Lucidité du corps. De l'empirisme transcendantal en phénoménologie* (Dordrecht: Kluwer Academic Publishers, 2001), 205–230.

105 Cf. Elizabeth A. Behnke, "On the interpretation and application of Husserl's research manuscripts" (Ferndale, WA: Study Project in Phenomenology of the Body, 2013).

106 Mohanty, *The Concept of Intentionality*, Part Three, Ch. 1; Elizabeth A. Behnke, "World without Opposite/Flesh of the World: A Carnal Introduction" (Felton, CA: California Center for Jean Gebser Studies, 1984; rpt. Study Project in Phenomenology of the Body, 1990; now available at www.lifwynnfoundation.org).

References

Behnke, Elizabeth A. "World without Opposite/Flesh of the World: A Carnal Introduction." Felton, CA: California Center for Jean Gebser Studies, 1984; rpt. Study Project in Phenomenology of the Body, 1990; now available at www.lifwynnfoundation.org.

— "On the interpretation and application of Husserl's research manuscripts." Ferndale, WA: Study Project in Phenomenology of the Body, 2013.

Brand, Gerd. *Welt, Ich und Zeit. Nach unveröffentlichten Manuskripten Edmund Husserls.* Den Haag: Martinus Nijhoff, 1955.

Broekman, Jan M. *Phänomenologie und Egologie. Faktisches und transzendentales ego bei Edmund Husserl.* Den Haag: Martinus Nijhoff, 1963.

Bruzina, Ronald. "Die Notizen Eugen Finks zur Umarbeitung von Edmund Husserls 'Cartesianischen Meditationen.'" *Husserl Studies* 6 (1989): 97–128.

— "Translator's Introduction." In Eugen Fink, *Sixth Cartesian Meditation: The Idea of a Transcendental Theory of Method, with textual notations by Edmund Husserl,* vii–xcii. Bloomington, IN: Indiana University Press, 1995.

— "Construction in Phenomenology." In *The Reach of Reflection: Issues for Phenomenology's Second Century*, edited by Steven Crowell, Lester Embree, and Samuel J. Julian, 46–71. Boca Raton, FL: Center for Advanced Research in Phenomenology, www.electronpress.com, 2001.

— "Eugen Fink and Maurice Merleau-Ponty: The Philosophical Lineage in Phenomenology." In *Merleau-Ponty's Reading of Husserl*, edited by Ted Toadvine and Lester Embree, 173–200. Dordrecht: Kluwer Academic Publishers, 2002.

— *Edmund Husserl and Eugen Fink: Beginnings and Ends in Phenomenology, 1928–1938.* New Haven, CT: Yale University Press, 2004.

— "Phenomenology and Cognitive Science: Moving Beyond the Paradigms. Review Essay: Jean Petitot, Francisco J. Varela, Bernard Pachoud, Jean-Michel Roy, eds. *Naturalizing Phenomenology: Issues in Contemporary Phenomenology and Cognitive Science.*" *Husserl Studies* 20 (2004): 43–88.

Cairns, Dorion. *Conversations with Husserl and Fink*, edited by Husserl-Archives. The Hague: Martinus Nijhoff, 1976.

Chokr, Nader N. "Mind, Consciousness, and Cognition: Phenomenology vs. cognitive science." *Husserl Studies* 9 (1992): 179–197.

Costa, Vincenzo. "Die Erfahrung des Anderen. Phänomenologie, Behaviorismus und Spiegelneuronen." *Husserl Studies* 24 (2008): 231–241.

Depraz, Natalie. "L'empiricité transcendantale de la phénoménologie." Ch. 5 in her *Lucidité du corps. De l'empirisme transcendantal en phénoménologie,* 205–30. Dordrecht: Kluwer Academic Publishers, 2001.

— "What about the *praxis* of Reduction? Between Husserl and Merleau-Ponty." In *Merleau-Ponty's Reading of Husserl*, edited by Ted Toadvine and Lester Embree, 115–125. Dordrecht: Kluwer Academic Publishers, 2002.

Fink, Eugen. "Die phänomenologische Philosophie Edmund Husserls in der gegenwärtigen Kritik." *Kant-Studien* 38 (1933): 321–383; rpt. separately Berlin: Pan-Verlag, 1934; rpt. in his *Studien zur Phänomenologie 1930–1939*, 79–156. Den Haag: Martinus Nijhoff, 1966.

— "Das Problem der Phänomenologie Edmund Husserls." *Revue internationale de philosophie* 1 (1939): 226–270; rpt. in his *Studien zur Phänomenologie 1930–1939*, 179–223. Den Haag: Martinus Nijhoff, 1966.
— "L'analyse intentionnelle et le problème de la pensée spéculative." In *Problèmes actuels de la phénoménologie*, edited by Herman Leo Van Breda, 54–87 [German/French on facing pages]. Paris: Desclée de Brouwer, 1952.
— "Operative Begriffe in Husserls Phänomenologie." *Zeitschrift für philosophische Forschung* 11 (1957): 321–337.
— "Die Spätphilosophie Husserls in der Freiburger Zeit." In *Edmund Husserl 1859–1959. Recueil commémoratif publié à l'occasion du centenaire de la naissance du philosophe*, edited by Herman Leo Van Breda and Jacques Taminiaux, 99–115. La Haye: Martinus Nijhoff, 1959.
— *Studien zur Phänomenologie 1930–1939*. Den Haag: Martinus Nijhoff, 1966.
— *VI. Cartesianische Meditation. Teil 1: Die Idee einer transzendentalen Methodenlehre*, edited by Hans Ebeling, Jann Holl, and Guy van Kerckhoven; *Teil 2: Ergänzungsband*, edited by Guy van Kerckhoven. *Husserliana Dokumente* II. Dordrecht: Kluwer Academic Publishers, 1988.
— *Phänomenologische Werkstatt*, edited by Ronald Bruzina. 4 vols. Freiburg: Karl Alber, 2006ff.
Froese, Tom, and Shaun Gallagher. "Phenomenology and Artificial Life: Toward a Technological Supplementation of Phenomenological Methodology." *Husserl Studies* 26 (2010): 83–106.
Gabel, Gernot U. *Husserl. Ein Verzeichnis der Hochschulschriften aus westeuropäischen und nordamerikanischen Ländern 1912–1990*. Hürth: Edition Gemini, 1995.
Gadamer, Hans-Georg. "Die phänomenologische Bewegung." *Philosophische Rundschau* 11 (1963): 1–45.
Gallagher, Shaun. "Hyletic experience and the lived body." *Husserl Studies* 3 (1986): 131–166.
— "Lived Body and Environment." *Research in Phenomenology* 16 (1986): 139–170.
— "Body Image and Body Schema: A Conceptual Clarification." *Journal of Mind and Behavior* 7 (1986): 541–554.
— "Body schema and intentionality." In *The Body and the Self*, edited by J. Bermudez, A. Marcel, and N. Eilan, 225–244. Cambridge, MA: MIT Press, 1995.
— "Neurophilosophy and Neurophenomenology." In *Phenomenology 2005*, Vol. 5: *Selected Essays from North America*, edited by Lester Embree and Tom Nenon, 293–316. Bucharest: Zeta Books, 2007.
— "Cognitive Science." In *The Routledge Companion to Phenomenology*, edited by Sebastian Luft and Søren Overgaard, 574–585. London: Routledge, 2012.
— and Francisco Varela. "Redrawing the Map and Resetting the Time: Phenomenology and the Cognitive Sciences." In *The Reach of Reflection: Issues for Phenomenology's Second Century*, edited by Steven Crowell, Lester Embree, and Samuel J. Julian, 17–45. Boca Raton, FL: Center for Advanced Research in Phenomenology, www.electronpress.com, 2001.
Geniusas, Saulius. "The Post-War Reception of *Ideen I* and Reflection." In *Husserl's Ideen*, edited by Lester Embree and Thomas Nenon, 399–414. Dordrecht: Springer, 2013.

Heidegger, Martin. *Der Satz vom Grund.* Pfullingen: Neske, 1957.
Held, Klaus. *Lebendige Gegenwart. Die Frage nach der Seinsweise des transzendentalen Ich bei Edmund Husserl, entwickelt am Leitfaden der Zeitproblematik.* Den Haag: Martinus Nijhoff, 1966.
Hoche, Hans-Ulrich. *Handlung, Bewußtsein und Leib. Vorstudien zu einer rein noematischen Phänomenologie.* Freiburg: Karl Alber, 1973.
Hoffmann, Gisbert. "Die Zweideutigkeit der Reflexion als Wahrnehmung von Anonymität." *Husserl Studies* 14 (1997): 95–121.
Horst, Steven. "Phenomenology and psychophysics." *Phenomenology and the Cognitive Sciences* 4 (2005): 1–21.
Husserl, Edmund. "Vorlesungen zur Phänomenologie des inneren Zeitbewußtseins." *Jahrbuch für Philosophie und phänomenologische Forschung* 9 (1928): viii–ix, 367–498, also Sonderdruck, Halle a.d. Salle: Max Niemeyer, 1928; rpt. in *Hua* X, 3–134.
— "Formale und transzendentale Logik. Versuch einer Kritik der logischen Vernunft." *Jahrbuch für Philosophie und phänomenologische Forschung* 10 (1929): v–xiii, 1–298, also Sonderdruck, Halle a.d. Salle: Max Niemeyer, 1929; rpt. in *Hua* XVII, v–xv, 5–335.
— "Die Krisis der europäischen Wissenschaften und die transzendentale Phänomenologie. Eine Einleitung in die phänomenologische Philosophie." *Philosophia* 1 (1936): 77–176; rpt. in *Hua* VI, 1–104.
— *Erfahrung und Urteil. Untersuchungen zur Genealogie der Logik*, edited by Ludwig Landgrebe. Prag: Academia-Verlag, 1939; rpt. Hamburg: Claassen & Goverts, 1948; rpt. Hamburg: Felix Meiner, 1999.
— *Die Krisis der europäischen Wissenschaften und die transzendentale Phänomenologie. Eine Einleitung in die phänomenologische Philosophie*, edited by Walter Biemel. *Husserliana* VI. Den Haag: Martinus Nijhoff, 1954.
— *Erste Philosophie (1923/24). Zweiter Teil. Theorie der phänomenologischen Reduktion*, edited by Rudolf Boehm. *Husserliana* VIII. Den Haag: Martinus Nijhoff, 1959.
— *Zur Phänomenologie des inneren Zeitbewußtseins (1893–1917)*, edited by Rudolf Boehm. *Husserliana* X. Den Haag: Martinus Nijhoff, 1966.
— *Formale und transzendentale Logik. Versuch einer Kritik der logischen Vernunft. Mit ergänzenden Texten*, edited by Paul Janssen. *Husserliana* XVII. Den Haag: Martinus Nijhoff, 1974.
— *Zur phänomenologischen Reduktion. Texte aus dem Nachlass (1926–1935)*, edited by Sebastian Luft. *Husserliana* XXXIV. Dordrecht: Kluwer Academic Publishers, 2002.
— *Die Lebenswelt. Auslegungen der vorgegebenen Welt und ihrer Konstitution. Texte aus dem Nachlass (1916–1937)*, edited by Rochus Sowa. *Husserliana* XXXIX. Dordrecht: Springer, 2008.
— *Späte Texte über Zeitkonstitution (1929–1934). Die C-Manuskripte*, edited by Dieter Lohmar. *Husserliana Materialien* VIII. Dordrecht: Springer, 2006.
Kersten, Fred. "Notes from the Underground: Merleau-Ponty and Husserl's Sixth Cartesian Meditation." In *The Prism of the Self: Philosophical Essays in Honor of Maurice Natanson*, edited by Steven Galt Crowell, 43–58. Dordrecht: Kluwer Academic Publishers, 1995.
Kockelmans, Joseph J., ed. *Phenomenology: The Philosophy of Edmund Husserl and Its Interpretation.* Garden City, NY: Doubleday, Anchor Books, 1967.

Kortooms, Antoine. "Fungerende intentionaliteit bij Husserl." In *Gehelen en fragmenten. De vele gezichten van de filosofie*, edited by Bart Raeymaekers, 124–129. Leuven: Universitaire pers Leuven, 1993.

Landgrebe, Ludwig. "Husserls Abschied vom Cartesianismus." *Philosophische Rundschau* 9 (1962): 133–177; rpt. in his *Der Weg der Phänomenologie. Das Probleme einer ursprünglichen Erfahrung,* 163–206. Gütersloh: Gerd Mohn, 1963.

Luft, Sebastian. *"Phänomenologie der Phänomenologie." Systematik und Methodologie der Phänomenologie in der Auseinandersetzung zwischen Husserl und Fink.* Dordrecht: Kluwer Academic Publishers, 2002.

Melle, Ullrich. *Das Wahrnehmungsproblem und seine Verwandlung in phänomenologischer Einstellung. Untersuchungen zu den phänomenologischen Wahrnehmungstheorien von Husserl, Gurwitsch und Merleau-Ponty.* Den Haag: Martinus Nijhoff, 1982.

Merleau-Ponty, Maurice. *Phénoménologie de la perception.* Paris: Gallimard, 1945.

— "Le philosophe et son ombre." In *Edmund Husserl 1859–1959. Recueil commémoratif publié à l'occasion du centenaire de la naissance du philosophe*, edited by Herman Leo Van Breda and Jacques Taminiaux, 195–220. La Haye: Martinus Nijhoff, 1959.

— *L'Œil et l'Esprit.* Paris: Gallimard, 1964.

— *Le visible et l'invisible*, edited by Claude Lefort. Paris: Gallimard, 1964.

— *Notes de cours 1959–1961*, edited by Stéphanie Ménasé. Paris: Gallimard, 1996.

— "Husserl aux limites de la phénoménologie. 1960," edited by Franck Robert. In Merleau-Ponty, *Notes de cours sur L'Origine de la géométrie,* 11–92, suivi de *Recherches sur la phénoménologie de Merleau-Ponty,* sous la direction de Renaud Barbaras. Paris: Presses Universitaires de France, 1998.

Mohanty, J. N. "Husserl's Concept of Intentionality." *Analecta Husserliana* 1 (1971): 100–132.

— *The Concept of Intentionality.* St. Louis: Warren H. Green, 1972.

Mohanty, J. N. "Understanding Husserl's Transcendental Phenomenology: An Introductory Essay." In *Apriori and World: European Contributions to Husserlian Phenomenology,* edited and translated by William McKenna, Robert M. Harlan, and Laurence E. Winters, 1–20. The Hague: Martinus Nijhoff, 1981.

Petit, Jean-Luc. "On the relation between recent neurobiological data on perception (and action) and the Husserlian theory of constitution." *Phenomenology and the Cognitive Sciences* 2 (2003): 281–298.

Seebohm, Thomas. *Die Bedingungen der Möglichkeit der Transzendental-philosophie. Edmund Husserls transzendental-phänomenologischer Ansatz, dargestellt im Anschluss an seine Kant-Kritik.* Bonn: Bouvier, 1962.

— *Hermeneutics: Method and Methodology.* Dordrecht: Kluwer Academic Publishers, 2004.

Spileers, Steven, comp. *Edmund Husserl Bibliography. Husserliana Dokumente* IV. Dordrecht: Kluwer Academic Publishers, 1999.

Szilasi, Wilhelm. *Einführung in die Phänomenologie Edmund Husserls.* Tübingen: Max Niemeyer, 1959.

Toadvine, Ted, and Lester Embree, eds. *Merleau-Ponty's Reading of Husserl.* Dordrecht: Kluwer Academic Publishers, 2002.

Van Breda, Herman Leo. "Le sauvetage de l'héritage husserlien et la fondation des Archives-Husserl"/"Die Rettung von Husserls Nachlass und die Gründung des Husserl-Archivs." Trans. Rudolf Boehm. In *Husserl et la pensée moderne/Husserl und das Denken der Neuzeit*, edited by Herman Leo Van Breda and Jacques Taminiaux, 1–42/43–77. La Haye/Den Haag: Martinus Nijhoff, 1959.

— "Maurice Merleau-Ponty et les Archives-Husserl à Louvain." *Revue de métaphysique et de morale* 67 (1962): 410–430.

van Kerckhoven, Guy. "Die 'Idee' einer transzendentalen Methodenlehre in phänomenologischer Absicht. Eine Voranzeige zu Eugen Finks Entwurf einer VI. Meditation zu Husserls *Méditations cartésiennes.*" In *Phänomenologie im Widerstreit. Zum 50. Todestag Edmund Husserls*, edited by Christoph Jamme and Otto Pöggeler, 81–105. Frankfurt am Main: Suhrkamp, 1989.

— *Mundanisierung und Individuation bei E. Husserl und E. Fink. Die VI. Cartesianische Meditation und ihr "Einsatz."* Würzberg: Königshausen & Neumann, 2003.

Varela, Francisco J. "The Naturalization of Phenomenology as the Transcendence of Nature: Searching for generative mutual constraints." *Alter* 5 (1997): 355–381.

Waldenfels, Bernhard. *Phänomenologie in Frankreich.* Frankfurt am Main: Suhrkamp, 1987.

— *Einführung in die Phänomenologie.* München: Wilhelm Fink, 1992.

Zahavi, Dan. *Self-Awareness and Alterity: A Phenomenological Investigation.* Evanston, IL: Northwestern University Press, 1999.

— "Review Article: Thompson, Evan. *Mind in Life: Biology, Phenomenology, and the Sciences of Mind.*" *Husserl Studies* 25 (2009): 159–168.

—

Elizabeth A. Behnke
Study Project in Phenomenology of the Body
PO Box 66
Ferndale, Washington 98248-0066
USA
sppb@openaccess.org

Recollection, Perceptual Phantasy, and Memorials: A Husserlian Proposal

Javier Enrique Carreño

Abstract

Throughout Edmund Husserl's writings on phantasy, memory, and image-consciousness one can find a close link between recollection and images, not only because for a time Husserl thought that every awareness of the past had an imagistic structure, but also because he repeatedly found images capable of conveying time and, especially, a past time. One would have expected Husserl to abandon the thesis of the past in images the moment he realizes that the hard core of recollection is not imagistic. But the opposite is the case. As Husserl arrives at a non-imagistic understanding of memory, he also develops a non-depictive account of images in terms of *perceptual phantasy*, together with a finer articulation of how images present us with the past. In this essay I show how Husserl's investigations into recollection and image consciousness enable us to speak of *memorials*, that is, images which not only prompt remembrance but which in point of fact reconnect a community with its past and with its destiny.

Keywords: Husserl, Phantasy, Memory, Memorials, Image-Consciousness

Proverbially we identify the past with what is no longer present, a moment of time that has vanished and 'gone out.' To ask for the whereabouts of time past, though, is not like fancying what the flame of a candle looks like after it is blown out. For Edmund Husserl, the past is consciously preserved, first in the retentional flow of time-consciousness, and then retrieved "once again" in recollection. Retention and recollection are the two principal modes in which we are aware of the past. And yet Husserlian phenomenology might still suggest other ways in which we become aware of the past in the present.[1]

Time and again Husserl suggests that a past time can appear in an image. In this essay we will show how this suggestion can connect to an account of *memorials*—a class of cultural objects ranging from public monuments to ritual and seasonal communal celebrations, whether secular or religious.

To unfold this suggestion we will first take stock of the marginal gains in Husserl's failure to assimilate recollection to pictorializations (§1). Following

1 See, for example, Christian Ferencz-Flatz, "Objects with a past: Husserl on 'ad-memorizing apperceptions,'" *Continental Philosophy Review* 45 (2012): 171-188, which analyzes Husserl's references to ad-memorizing apperceptions mainly in the context of Husserl's later phenomenology of association. My approach to Husserl's text differs from his in that I am concerned with memorial experiences that have images as their basis.

this failure, Husserl will not only formulate a *new* account of recollection, but also, he will broaden his understanding of images and our interaction with them from pictorialization to fictionalization (§2). These revisions will enable us to ask anew how the past can become present in an image via what may be called perceptual memory (§3).

§1. Recollection and pictorialization

Thinking philosophically about *recollection* [*Wiedererinnerung*] makes the ineluctable demand that we grasp how it differs from *perception* as originary consciousness. This philosophical demand to think of identity and difference stems above all from the phenomena themselves. I recollect, e.g., riding on a steam locomotive with my children through the Ötscher mountains. While I remember these events – the whistle blowing, the steam coming profusely from the engine, my daughter's surprise and my son's excitement as the train began to roll–they pass "before me" *once again* [*wieder*] as it were. I do not merely recall the occurence a series of mental vignettes. Rather, I reanimate the past yet again, inwardly and yet not just 'mentally.' For I witness the past from a quasi-embodied standpoint, indeed as if I perceived these events again—but without doubting that they are *past* and not present.[2]

That we are seldom (if ever) in doubt about the absence and above all the "pastness" of what we recollect cannot be explained simply by affirming, *ex hypothesi*, that the past no longer exists. Essential differences must be found on the level of how past and present each come to presentation (*Vorstellung*). For Franz Brentano, for example, the protean character of phantasms mark such phantasy and memorial presentations as surrogate (as opposed to direct) and conceptual (as opposed to full-fledged intuitive).[3] Husserl's approach, in contrast, is to up-

---

2 Edmund Husserl, *Husserliana 23. Phantasie, Bildbewusstsein, Erinnerung. Zur Phänomenologie der anschaulichen Vergegenwärtigung. Texte aus dem Nachlass: 1891–1925*, ed. Eduard Marbach (Den Haag: Martinus Nijhoff, 1980); English translation: *Collected Works XI. Phantasy, Image Consciousness, and Memory (1898–1925)*, trans. John B. Brough, (Dordrecht: Springer, 2005); henceforth abbreviated as *Hua* XXIII with German page reference (also indicated in the English translation). Throughout this paper I cite from this translation with only minor changes. Here see "Text Nr. 11." See also Edmund Husserl, *Husserliana 13. Zur Phänomenologie der Intersubjektivität. Texte aus dem Nachlass. Erster Teil: 1905-1920*, ed. Iso Kern (The Hague: Martinus Nijhoff, 1973), "Text Nr. 10."

3 See F. Brentano, *Grundzüge der Ästhetik*, ed. F. Mayer-Hillebrand. (Bern: Francke, 1959), 86–87. For Brentano, phantasy presentations are not intuitive, like perceptual

hold the genuineness and intuitiveness of memory, expectation, and phantasy. While remembering the train journey, the latter is not merely emptily or conceptually intended: rather, it hovers (*vorschwebt*) before me and 'happens' yet again: I 'feel' the regular forward movement of the train; I 'see' my children peering through the window—not as if I were now perceiving all these things, but rather as if they were "hidden by a veil" (*Hua* XXIII, 287) as absent from the present and in the mode of something past.

But Husserl's vindication of memory and of phantasy naturally reintroduces at another level the problem of the difference between perception and recollection. Setting aside genetic and empirical issues such as the intensity of phantasms, how is recollection altogether different from an act whose object is taken to be actually and corporeally present before me? If my recollection of the whistle blowing is exceptionally vivid, how does it differ from actually hearing a whistle blowing at a distance?

Around the period of his *Logical Investigations*, Husserl met this challenge by appealing to the well known "Apprehension and Content for an Apprehension" schema.[4] Perceptions, abstracted from their qualitative moments (e.g., affirming or doubting the existence of what is presented), share in the same "presentation" (*Vorstellung*) and, more specifically, in the same form of "perceptual apprehension" (*Wahrnehmungsauffassung*, cf. *Hua* XXIII, 5). The form of apprehension refers to how an immanently present sensuous content is 'taken up' or 'apprehended' in the presentation of something appearing as present (*gegenwärtig*). Thus, e.g., the colour-content "yellow" is taken up in the apprehension of, e.g., a sunflower before me. Phantasy and recollection, in contrast, would have their own distinctive "phantasy presentation" as their basis, accomplished by a phantasy apprehension (9). In the latter case, a phantasm would be 'taken up' in the appearing of something as absent and merely presentified (*vergegenwärtigt*, 16).

The application of the schema to recollection and phantasy does not yet explain how, e.g., a tone-phantasm enjoys the good fortune of being apprehended in my reminiscing and not in my perceiving of the whistle blowing (cf. *Hua* XIX/2, 623). To tackle this problem, Husserl tentatively forwards *pictorialization* (*Verbildlichung*) as the distinctive characteristic of all presentifications (*Vergegenwärtigungen*), i.e., acts like recollection, phantasy, and empathy,

presentations, but still enjoy an intuitive 'core; whereby they approximate perceptual presentations.

4 Edmund Husserl, *Husserliana 19/1 und 19/2. Logische Untersuchungen. Zweiter Teil. Untersuchungen zur Phänomenologie und Theorie der Erkenntnis*, ed. Ursula Panzer (The Hague: Martinus Nijhoff, 1984), henceforth abbreviated as *Hua* XIX/1 and XIX/2 and page reference, respectively. Here *Hua* XIX/2, 621ff.

where an object is intended intuitively but not given as presently and corporeally before me. According to this suggestion, my recollecting of the train ride would be analogous to my revisiting of this event through physical images. With this suggestion Husserl does not mean, however, that mental representations abide in the mind akin to how one naively takes external representations to abide in the world. Even an external image, for Husserl, is already an intentional accomplishment. Such an image is neither a *thing* nor an inherent part or property of anything, but rather, it is an *object*, constituted by consciousness on the basis of a physical appearance so as to represent something else by way of resemblance.[5]

However, the hypothesis that recollections are pictorializations still undermines the aforementioned straightforwardness of the past, and renews a rather limited pairing of memory to physical images. Even St. Augustine, who speaks of memory as composed of "fields and vast palaces (...) where are stored the innumerable images of material things," also struggled to comprehend how memory relates—without images—to one's own past *self* and mind (including prior remembrance and forgetting); abstract principles and laws; and God.[6] It is not surprising that Husserl himself would initially harbor reservations towards his own pictorialization hypothesis[7] and that a more careful analysis of pictorialization in external images would in fact help him surrender it.[8]

Husserl's most extensive presentation of image-consciousness appears in the third part of a lecture course entitled *The Principal Parts of Phenomenology and Theory of Knowledge* from the Winter semester of 1904–05. In this course, Husserl suggests that an external image entails three distinct but interwoven moments: 1) the *physical image thing* or material object that serves as a substratum to 2) the *image-object*, a figment appearance with a determinate form and coloration. The image-object is endowed with the function of representing by resemblance 3) the *image-subject, that* which appears not in its own (perceptual) ap-

5 The critique of the naïve "representative theory of consciousness" (*Bildertheorie*) in the *Logical Investigations* remains in this regard a lasting achievement. See *Hua* XIX/1, 436–440.

6 Augustine, *Conf.* X, 8–27, trans. F.J. Sheed (Indianapolis: Hackett, 1993).

7 Cf. *Hua* XXIII 16fn. and 617. Also 201 (*ca.* 1904): "Reid believes that the memorial presentation is an immediate presentation. Is there not in fact a great difference between a presentation that is consciously pictorial, such as a presentation by means of genuine images (actual image apperception), and presentations by means of *phantasy images* and *memory images*?"

8 In texts such as *Hua* X N. 39 and *Hua* XXIII N. 8 from 1907 and 1909 respectively we see a clear break from the pictorialization thesis vis-à-vis retention (primary memory) and phantasy. The thesis of retention as pictorialization comes up again in the Bernau Manuscripts, but not sustained without objections. See Toine Kortooms, *Phenomenology of Time. Edmund Husserl's Analysis of Time-Consciousness* (Dordrecht: Kluwer, 2002).

pearance but precisely *in* image. Thus, a photograph of my child consists not just of glossy photographic paper [=1]. For as the latter triggers my imagining on its sensuous basis of an "image child" [=2], "my child" [=3] appears photographically in it (cf. *Hua* XXIII 20).

Already from this three-fold distinction we gather that simple perception cannot account for external images, since a presently perceived object cannot, in and of itself, represent an *absent* object without the mediation of an *unreal* object. And so, only a double movement of apprehension can account for such mediation. First, an *image-object* apprehension takes up the sensuous contents corresponding to the perceptual presentation of a physical thing so as to produce an image appearance. Effectively, when beholding an image, I no longer "see" canvas and paint brush strokes merely for themselves and apart form the image appearance that emerges on their sensuous basis. Second, (and in this sense differently from signitive apprehensions), a concurrent *image-subject* apprehension pervades the image-object internally with the new "apprehension-characteristic" of making the intended subject intuitable. This means that when I behold an image, I do not see an image-object merely by itself, since the latter lends appearance to the *image-subject* that thus appears "*in* image."

Tallying the image and subject apprehensions together, we could speak of a dual displacement in image-consciousness: first, of the sensuous contents quarried by the image from a real appearance so as to lend appearance to something else; and second, of the subject (*Sujet*) which appears precisely *in* that borrowed appearance and displaced from its own original appearance. Such displacements prevent the positing of the unreal image-subject as being currently and presently before us, and in this sense they create a permanent distance between what is really present and what is fictive. But it is no less true that precisely such distance enables something fictive to appear *in* the real world precisely *as* fictive. The image does not just wall the image-subject off from reality, but like a window pane, it shows it through and lets it come to appearance as absent. This is perhaps the strong point of Husserl's analysis of image consciousness: as soon as I apprehend an image, I *see* the subject. And while the subject appears mediately—in image—*what* appears is nevertheless *the* intended subject and not another likeness. Indeed, when turning towards a photo of my child, I do not see something that more or less resembles my child: rather, I see *my* child, rendered more or less accurately:

> What functions representatively in the content of the image object is conspicuous in a specific way: it *exhibits*, it *represents*, pictorializes, *makes intuitable*. The subject looks at us, as it were, through *these* traits (30).

And yet, the image subject appears unmistakably as *absent* because the image as a whole has at least two different kinds of limits with the *present*. One of them is the *point* at which an image's resemblance with what it depicts would be so complete so as to impinge the subject on our conscious regard as if it were something presently perceived. Were the image to reach this point (e.g., in a convincing life-size cutout, in a compelling wax figure, or in an animatronics display) it would precisely foil the consciousness of an image. Another limit is the *frame* or border setting the image apart from its actual surroundings, and in the absence of which I would have to posit the image as something perceived (e.g., a life-size cutout seamlessly blending with its surroundings and deterring an image apprehension). The perceptual regard posits what it grasps as being, whereas within the framed space of the image no such *thing* appears: literally, what appears is a perceptual *no-thing* (*ein Nichts*) in the sense of an ideal object.[9]

So at once the image is not *quite* like the things that surround it, and also *not quite* 'like' what it represents. Husserl will describe this double inadequacy of the image as a consciousness of conflict that cannot be resolved by nullification and a new positing perception. In other words, an image is not demoted to a merely illusory figment, because an image is precisely the kind of object that announces its unreality to our gaze from the start and beckons our imagination to come into play and take it for an image *of* something absent.

Thanks to these two limits—which we may call the image's *horizontal* limit with the world and the *transversal* distance between image and subject—Husserl may speak of images as imaginary niches of unreal space and time, carved as it were *in* the midst of real space and time:

> So we have appearance here, sensuous intuition and objectification but in *conflict* with an experienced present. We have the appearance of a not now *in the now*. '*In* the now,' insofar as the image-apprehending is something temporally now. Yet on the other hand a '*not now*' insofar as the conflict makes the image object into a *nullity* that does indeed appear but is *nothing*, and that may serve to exhibit something existing (*Hua* XXIII 47–48).

9 Cf. *Hua* XXIII, 46 and 537. Husserl will also affirm the perceptual "no-thingness" and "ideality" of the image-subject (in 537), even though this affirmation seems at first hard to reconcile with Husserl's statements that, e.g., the subject of a photograph of the Berlin palace *is* the palace *in* Berlin, or that, when I turn to a photo of my child I see not an image-child but *my* child. Still, the image subject has no independent appearance apart from the image: it is precisely a moment of image-consciousness. In another registry one would say that the image-subject is the *motif* for an image, an identity constituted in a manifold of images.

The act of image-apprehending is *nunc*, and yet *what* appears in an image before me is not *nunc*. But could the '*not now*' in an image refer to an *absent* time, and more specifically, to a '*no longer now*'? For in that case, pictorialization could analogously account for our awareness of the past in recollection. This is the suggestion that Husserl pushes near to its breaking point in a text *ca* 1904:[10]

> The painting furnishes a perceptual presentation, but what is now is the representant of something that is not now. Of course, not every painting represents a time, but think of paintings of historical events; the execution of Charles II, the storming of the Bastille, and the like. Yet these are not memories (*Hua* X, 183).

For recollection does not intend the past object *in* an image-appearance that analogizes or resembles the past to a degree. Rather, recollection intends the past object itself in and through its *own* reproduced, past appearance (Cf. *Hua* X, 183). In other words, there is no *distance* in recollection between an *image-object* appearance and an appearing *image-subject*.[11] Moreover, in recollection, the memorial appearance has no boundaries or limits with the perceptual field of regard (Cf. *Hua* XXIII, 49). The absence of limits and of conflict with perceptual reality here makes it impossible to justify a double apprehension of both image *and* subject in recollection. Furthermore, the "*not now*" presented in an image does not *eo ipso* have the characteristic of a 'sunken' temporal phase vis-à-vis the present and continuously receding from the present. Such difficulties will lead Husserl to altogether discard the model of pictorial-ization as a plausible account for recollection.

That internal pictorializations do not account for recollection would seem to suggest that external pictorializations also cannot have anything to do with our awareness of the past. And yet Husserl does not give up the idea that images can and do refer to the past. At the very least one can say of an image-appearance that it can invite a transition to recollections. Literally, images "bring the past to mind" (*erinnern*) in a new presentation:

> A photograph, when it is particularly good, re-presents a person to us. We immerse ourselves visually in the photograph. A photograph, however, can also bring to mind a person in a manner similar to that in which a sign brings to mind something signified. If it does that, the image is characterized, phenomenologically, as that which

10 See Rudolf Bernet's "Einleitung," in: Edmund Husserl, *Texte zur Phänomenologie des inneren Zeitbewusstseins (1893–1917)*, ed. Rudolf Bernet (Hamburg: Felix Meiner Verlag, 1985), xviii–xix.

11 As John Brough puts it, "the backbone of Husserl's critique of the image theory is that it cannot explain memory as a direct consciousness of the past." See his "Husserl on Memory," In *Edmund Husserl: Critical Assessments of Leading Philosophers*, ed. Rudolf Bernet, Donn Welton, and Gina Zavota (London: Routledge, 2005), 334.

> brings something to mind. The person himself, however, who exists for us intentionally in a second and separate, though related, presentation—for example, in a phantasy presentation (but perhaps also in a merely empty intention)—the person, I say, then appears as what the image brings to mind (52).[12]

Not only photographs but also artworks may externally symbolize a past time and trigger phantasy presentations of that past:

> When the image operates aesthetically, it may indeed be the case that a new presentation brings the subject or some of its components to a more complete intuition—say, to a more fitting coloration. Universally, the play of phantasy may be set in motion in such a way that we become immersed in the world of the subject, as when, on seeing the pictures of Paolo Veronese, we feel ourselves transplanted into the magnificent, opulent life and activity of the grand Venetians of the sixteenth century; or as when we see in Dürer's agreeable woodcuts the transfiguration of the German landscape and the German people of his time (37).

And perhaps photographic and cinematic images come closest to not only symbolizing but in fact *internally analogizing* a past duration. Images can and do portray events, and this is no less true of motion pictures than of so-called static images. As Husserl puts it in a text *ca.* 1912.

> Rest or change appears in the image. In the ordinary resting image, which depicts by means of an unchanging image object, a movement might appear—a rider galloping away in a painting, for example. In the mutoscope, however, an image object that is itself moving appears; and in that case movement is presented by movement, and so on (*Hua* XXIII, 489n).[13]

Photographic images represent a time "that-has-been," as Roland Barthes puts it,[14] even though "when" in the past is itself an empty intention, and so a memorial image is never quite free from symbolization and other determinations that obtain from the concrete image-making technique employed and the culturally conditioned, empirical usages of images.[15]

However, an image that brings the past to mind by depicting it and by prompting reminiscences and imaginings may be faulted with displacing us from the present. For Husserl, I cannot be equally present to a recollected past and to

12 See also *Hua* XXIII, 90.

13 Cf. Javier Carreño, "On the Temporality of Images according to Husserl." *The New Yearbook for Phenomenology and Phenomenological Philosophy* 8 (2008): 73–92.

14 Roland Barthes, *Camera Lucida*, trans. Richard Howard (New York: Hill and Wang, 1982), 96.

15 See, e.g., *Hua* XXIII, 493–494 and 496–497, where Husserl discusses the indeterminacy of the subject in images made after a description of something past or fictive.

the perceived present,[16] and so, in a sense, recollecting in this manner can distance us from the present. Our mindfulness about the present is diminished, and, as we dwell in remembrance, we become "absent-minded." The initial opposition between perception and memory (in the sense of recollection) would seem to re-establish itself again with regards to *memorial images*.

Still, one could object that such *memorial images* or images that prompt reminiscences are not quite *memorials* understood as images that re-enact the past and bring it *into* the present, in ways that matter *in* the present and (perhaps) *for* the future. Evidently, a closer determination of memorials in this sense requires that we rethink both the presence of the past and the presence of something in an image, and Husserl gives us a head-start on both in the analyses ensuing from the failure of the pictorialization hypothesis for recollection.

§2. Recollection and Perceptual Phantasy

That pictorialization is not the way in which we become aware of the past is not only a milestone in philosophically thinking about memory: it is also a formidable challenge for phenomenology, and it is hardly surprising that it took years for Husserl to take in its implications. For what this realization affirms is precisely that the past cannot appear in a memorial appearance or apprehension somehow distinct from an original perceptual appearance. When I recollect the ride on the Ötscherland Express with my children, these past events are not mediated to me via a non-perceptual appearance: for they are *directly* intuited— as directly as when I first perceived, e.g., the steam column, and the train setting into motion. As Husserl puts it, "Memory is a direct presentation of the past, just as perception is a direct presentation of what is present" (*Hua* XXIII, 235). Moreover, the past-ness of what I remember is nothing presented in addition to what is remembered, and certainly nothing that I decide on the basis of a mere presentation of a column of steam emerging from locomotive. The past-ness of what appears in memory is neither inferred from the appearance and its contents nor imposed on an appearance on some other basis.[17] Whether what appears is present or absent, past, present (but not perceived), future, or fictive, has to do rather with the *manner* in which this act of presentification is innerly or internally experienced in the absolute flow of consciousness.

16 Cf. *Hua* XXIII, 75.

17 Cf. *Hua* XXIII, 267: "Just as in perception we had the color adumbration as the apprehension content for the objective color (the color of the physical thing), so in phantasy we have the color adumbration as the apprehension content for the objective color. It is the same on both sides."

And an appearance is innerly experienced either as an *impression*, a current experience, or else, it is experienced as a *reproduction* of another impression. The sensuous content for a perceptual apprehension, for example, is not a raw datum of consciousness, something merely experienced, but rather, it is already a unity for which there is an impression in the absolute flow.[18] A current experience, such as an ongoing act of recollecting, is also a temporal unity "constituted" in the absolute flow. And yet, such an act of recollecting involves neither actual sensuous contents nor actual apprehensions[19] to intend something transcendent. In this regard, recollection cannot be confused with perception, for there is no simple, impressional consciousness of its contents in absolute consciousness. Instead, there is an awareness of these contents as otherwise than impressions, i.e., as reproductions of impressions. But again, by "reproduction" Husserl does not mean that in recollection "the originary experience belonging to the earlier internal consciousness is now being produced again, as if an echo, reflection, afterimage of the earlier internal consciousness, although weak, were coming back" (310). The recollection of something past does not produce a past appearance again, but rather reproduces the impressional consciousness that corresponds to the original appearance of that object. As such, the past, perceptual appearance is not actually produced, but only "intentionally implicated" by the reproduction of its corresponding, earlier *impressional consciousness*.[20]

Recollection is thus "a new kind of *act*" (310) even without its own unique apprehension form or unique sort of content. It is the direct presentification of something past that intentionally implicates the original appearance and contents of the corresponding past act via the reproduction of its impressional consciousness. This novel formulation of recollection is clearly different from that of external pictorialization (or mediated presentification). In perceiving a thing, only

18 Cf. *Hua* XXIII, 308–311.

19 "[A] re-presented [*vergegenwärtigter*] act is not something really immanent [*Reelles*] in internal consciousness" (309); "Concerning the 'phantasms,' however, it was naturally false to say of them (within the boundaries of internal consciousness) that they were 'experiences [*Erlebnisse*]';" for experience was equivalent to: 'A datum of internal consciousness, something internally perceived'" (310).

20 Husserl does not conceal his dissatisfaction with the couplet "impression" and "reproduction"—terms borrowed from an Empiricist tradition, which, in this context, easily trigger the wrong associations. An impression is not the impact of a sensuous content on consciousness, but the absolute consciousness of an original appearing of that content. Reproduction, in turn, is not the rendering of a sensuous appearance—akin to pictorialization—but the reproduction of (absolute) consciousness, and more specifically, in recollection, the retrieval of an earlier impressional consciousness, such that, when I remember, I am not aware of currently perceiving something (past) but only of *having perceived* that past object.

one object is intended, and it appears in its own appearance. In pictorialization, the intended object is apprehended in an image appearance. In recollection, the intended object appears directly in an original appearance, but this appearance corresponds to and is nested in a *reproduced*, past flow of consciousness. We have said of pictorialization that a present object could not depict something absent without the mediation of an *unreal* object. Of recollection we now have to say in turn that consciousness could not directly face something *past* without *virtualizing* itself so as to virtually stand on its own 'earlier' flow.[21]

This new account of recollection *ca.* 1909 broke new ground for Husserl's entire understanding of presentifications, and probably also for his account of time-consciousness. Henceforth Husserl could account for the peculiar differences between memory and expectation, empathy, and pure phantasy, by reference to the flow that is reproduced in each case, e.g., earlier as opposed to forthcoming, other as opposed to proper, or else wholly inactual. Basic issues such as whether something appears present or absent were no longer made dependant on empirical traits such as the vivacity of the contents, but were now referred to the *manner* in which absolute consciousness internally and pre-egoically locates its own experiences. Husserl could also now articulate how recollection (and, *a fortiori*, phantasy), while distinct from perception, nevertheless presents the same object "in the same way"—a basic requirement for launching phenomenology as an eidetic science.[22] Furthermore, vexing problems of iterability, e.g., how there can be memory of a memory, memory of a phantasy, phantasy of a memory, memory of an external image, and so on, could now credibly be sorted out by distinguishing between the impression corresponding to a reproductive act and the reproduction of an impressional consciousness.

Last but not least, with this new account of recollection, Husserl can now better fathom the limits of my past and of 'my lifetime.' There are limits to what I can remember, for example, I can only go back so far in my memory. In trying to remember the time before my childhood I run into a darkening of the memorial field, but not as if, e.g., I arrived at the very bottom of a cave. For every point in my memorial nexus motivates the constitution of a co-existing world, as well as of a time before that time, and even a time *before* my birth. And the possibility of having perceived such a time stands not merely as an empty or ideal

21 Husserl does not use the term 'virtuality' in this context, but I credit Nicolas de Warren for speaking about the virtuality of the past at a talk at the Cologne-Louvain Summer School in Phenomenology at the Husserl Archives of Cologne (June, 2012).

22 See Julia Jansen, "Phantasy's Systematic Place in Husserl's Work. On the Condition of Possibility for a Phenomenology of Experience." In *Edmund Husserl: Critical Assessments of Leading Philosophers*, ed. Rudolf Bernet, Donn Welton, and Gina Zavota (London: Routledge, 2005), 221–243.

possibility to me, but precisely as a possibility motivated by *my* remembering. I do not ask what the unperceived past would have looked like to an anonymous perceiver, but rather, I ask, how such past would have appeared (to me), had *I* lived then.[23]

This last point brings us back to *memorial images*. By now it is clear that, if at least some images refer us to the past, this is neither because recollection pictorializes nor because some pictorializations perfectly analogize the past. But my recollection can and does motivate the belief that even an unperceived past *could* have appeared to *me*. Would not recollection likewise motivate me to find an appearance of the past in an image?

> Something not present (something that in other circumstances would be intuitive and even be presented in a reproduction or else in perception) is pictorialized, and rendered perceptible to the senses for me in the perceptual figment. The figment masks from me the presentifying (reproductive) presentation, coincides with it: what is presentified slides into what is present, which turns into what is exhibiting (*Hua* XXIII, 383–384).

Instead of reproducing an absent, 'earlier' flow so as to intentionally implicate a past perception, the very opposite can also occur, namely, that I intentionally implicate a present perception in the recollection of something past, such that the past appears in image analogous to how I would have recollected it. Consciousness can, as it were, virtualize its own *current* positing perception (in the sense of *Wahrnehmung*) so as to merely perceive (in the sense of *Perzeption*) something intuitively given as absent.

In this reformulation of image-consciousness in terms of *perceptual phantasy*, inspired by the new analysis of recollection, and completed *ca.* 1918[24] there are already three major challenges to the earlier account of external pictorialization.

First, in our awareness of images, a *double* apprehension—of an image *and* of a subject—is no longer required for having image-consciousness, since the modified perception (*Perzeption*) is a direct and not a mediate apprehension. This realization is a challenge for Husserl, not because it runs counter to the earlier analysis of image-consciousness as a double apprehension, but rather, because it goes *beyond* this analysis.[25] Admittedly images such as photographic portraits do awaken a double apprehension. But other images simply demand that we take what they exhibit as unreal in the first place. Briefly put, some images require not so much that we *see* something *else* in image, as that we *first*

---

23 Cf. *Hua* XIII, 295–296.

24 See especially Hua XXIII, Text N. 18b.

25 Cf. *Hua* XXIII, 471–472 and fn.

modify the way we experience "seeing" so as to "see" differently. Images that show an object by resemblance have the function of "depictiveness" (*Abbildlichkeit*), while images that primarily propose an imaginary mode of experiencing what appears perceptually have the function of "imaging" (*Bildlichkeit*), i.e., of giving an unreal image "frame" to our experience.[26]

Second, according to this function of "imaging," the role of phantasy vis-à-vis external images also changes. In depictive image-consciousness phantasy produces an intuitive appearance of something absent by giving an image-apprehension to the sensuous contents of an otherwise positing perceptual apprehension. But in the function of "imaging," phantasy does not produce an intuitive image appearing. Rather, phantasy takes up an already intuitive perceptual appearance of an actual object into 'imaging.' But to say that "perceptual phantasy" is intuitive on the basis of a perception (*Perzeption*) and not on the basis of a phantasy (image) apprehension or of a pure phantasy reproduction means that the phantasy in "perceptual phantasy" is of a "non-intuitive" sort (*unanschaulicher*, cf. 514). Such a "non-intuitive" phantasy is a challenge for Husserl, but not because it forces him to abandon his initial stance on phantasy as an intuitive consciousness. The difficulty with "non-intuitive phantasy" is rather the suggestion that phantasy may have more to do with fictionalizing experiences than with merely bringing objects to intuition.

Third, this "imaging" function is not necessarily sustained by the consciousness of horizontal and transversal limits as we described earlier. The boundary between an image and reality is rather internally established by the free act of perceptual phantasy itself. But if this act seems less constrained by external conflict with reality, it is also less exposed to the relativism of, e.g., arbitrarily establishing the boundaries between reality and fiction, because it is far more dependent on extrinsic and empirical elements. At the very least an act of perceptual phantasy depends on a 'culturally informed' knowing of the concrete imaging form and rules of engagement giving rise to the modification of positing perception into perceptual phantasy, e.g., for the temporal arc of a play or show.

26 Cf. *Hua* XXIII, 515. According to the entry for "Bildlichkeit" in the *Deutsches Wörterbuch von Jacob und Wilhelm Grimm* (16 Bde. in 32 Teilbänden. Leipzig 1854–1961. Quellenverzeichnis Leipzig 1971. Online-Version: *Deutsches Wörterbuch von Jacob und Wilhelm Grimm*, Accessed July 17, 2013. http://woerterbuchnetz.de/DWB/?sigle=DWB&mode=Vernetzung&hitlist=&patternlist=&lemid=GA00011), the term refers to the *sensus allegoricus* or figurative meaning to be contrasted with the literal sense. From Husserl's text, however, it is clear that *Bildlichkeit* is not a symbolic intending on the basis of an inner imaging. By *Abbildlichkeit* an image lends appearance to something absent. But by *Bildlichkeit* an image gives an imaginary frame to what otherwise would be the appearance of something present.

In this regard Husserl speaks of perceptual phantasy being an "empirical phantasy-modification" (519). And without such empirical clues as to the appropriate manner of imaginatively engaging non-depictive images, I may fail to see what I am meant to see along with the image. Hence, a perceptual phantasy is closer to what Husserl calls "a *logically (intellectually) mediated semblance*" (147), which John Brough qualifies as a matter of "interpretation, a kind of looking that lets what is in the picture manifest itself so that the picture can achieve disclosure"[27]

It is no wonder, then, that a more sustained analysis of theatrical representation will serve an exemplary role for fathoming the limits of *perceptual phantasy*. When attending a theatrical representation, the actors do not indeed strike me as representations,[28] but rather they are enacting something fictive, and upon coming to the performance, I, as a viewer, expect no less of them.[29] Properly I do not *see* Hamlet *in* the given actor, rather, I *perceive* Hamlet while remaining all along aware that Hamlet does not exist before me: only the actors and props on stage do. That this underlying realization does not petrify my enjoyment of the theater like a Medusa head has to do with the fact that I am "playing along" *via* perceptual phantasy, in the manner that is appropriate to the art form of drama, for the duration of the show. This is the strong point in Husserl's analysis of perceptual phantasy. In drama the fictive person or object is *directly* intended, while the real person or object is concealed (*verdeckt*). But how concealed—especially if the concealed objectivities are *continuously before our perceptual regard*? The actor does not stand behind or beneath the appearing person of the drama, as if he were the background for which the character is the foreground, or a mere mask-bearer. Rather, the actor *is* that person *of* the drama: he imaginatively takes himself to *be* that person and appears perceptually before an audience that imaginatively takes him *for* that person.[30] The actor is concealed by the

27 John B. Brough, "Picturing Revisited: Picturing the Spiritual," In *The Truthful and The Good: Essays in Honor of Robert Sokolowski*, ed. John J. Drummond and James G. Hart (Dordrecht: Kluwer, 1996), 55.

28 "But here 'image of' does not signify depiction of [...] The actor's presentation is not a presentation in the sense in which we say of an image object that an image subject is presented in it. Neither the actor nor the image that is his performance for us is an image object in which another object, an actual or fictive image subject, is depicted" (515).

29 "In the case of the theatrical presentation, the play—this section of an illusory world—appears, but we do not begin with a normal perception. We do not begin with the thesis of the reality of what appears perceptually" (516).

30 As Jorge Luis Borges once put it, "the actor plays at being another before a gathering of people who play at taking him for that other person." J. L. Borges, "Everything and Nothing," In *Obras Completas* Vol II. (Buenos Aires: Emecé, 2005), 192. English trans-

frame of a communal, perceptual phantasy, veiled in perceptual phantasy—but the veil is translucent.[31] One could even say that the dramatic image does not wall off the actor as a mere substratum, but shows him through, while all along rendering his effective appearance *qua* actor ineffective for the purposes of the image.

In our previous section (=§1) we found that for Husserl the past can appear depicted in an image. Naturally, the past can also appear imaged in a perceptual phantasy, such as a realistic, historical drama, without thereby ceasing to be mere fictionalization:

> The intent of realism: To present landscapes, human beings, human communities, destinies, and the interweavings of destinies, in the fullest possible 'characteristic' concreteness, as if we were seeing them, and, within a fixed frame, witnessing everything related to them in the richest possible fullness and in the substance of their being according to their innermost, though intuitive, motivations. What is characteristic. Characterizing a situation in a time and the time itself, the level of culture, the sort of life and life-form of this quarter of the globe, of this city, and so on (541).

One may say that realistic art forms and manifestations, which we engage through perceptual phantasy, have the capacity to express the spirit of an epoch or of a people with an intensity and concreteness that already affords aesthetic delight. This delight comes about *without* the artist having to resort to an "idealization" of the past and without having to substitute real types with ideal types. We delight in seeing the past, even a stark past, because at the very least we have here a "delight in seeing that understands."[32]

---

lation: J. L. Borges, *Labyrinths: Selected Short Stories and Other Writings*, ed. D. A. Yates and J. E. Irby (New York: New directions, 1962), 248.

31 "[W]e do not alternate intuitively between reality and illusion in such a way that one of the intuited things in the alternating is so to speak, concealed for precisely as long as the other is actually intuited (which is what happens in every conflict within the sphere of intuition). Rather, without alternating, we have from the beginning only the artistic "image"; and what is real that functions as presentation, what is actually experienced without modification is continuously concealed—*concealed though there is consciousness of it only consciousness of it nonintuitively* and in the peculiar fashion that the word concealment suggests in this case" (516–517, my emphasis).

32 Incidentally, the claim that the Aesthetic delight of certain forms of representation is precisely 'theoretical' is defended by Aristotle in *Poetics,* Ch. 4: "Though the objects themselves may be painful to see, we delight to view the most realistic representations of them in art, the forms for example of the lowest animals and of dead bodies. The explanation is to be found in a further fact: to be learning something is the greatest of pleasures not only to the philosopher but also to the rest of mankind [...] the reason of the delight in seeing the picture is that one is at the same time learning—gathering the meaning of things"

> Parallels with biography and historical characterization (individual characterization) of a time, of the people belonging to it, and so on. Such characterization is part of scientific cognition. Nothing is made up there; on the contrary, everything is described. Realistic art is also a sort of biography of a time, of the strata of a time. It portrays through characteristic 'images.' It constructs fictions in which characteristic types belonging to the time present themselves. It is art, not science, though in its own way it does mediate knowledge. It produces formations in the manner peculiar to phantasy, and as types with respect to times and world epochs (*Hua* XXIII, 541).

Art can come close to a "scientific" cognition of the past, and run parallel to it, yet without intersecting it: for realistic art is still "myth," not "fact." Still, such realistic dramas are not, for that reason, less relevant or meaningful for an individual or a community to experience a time past. For fiction does mediate a form of knowledge, at the very least a knowledge of types, but also—and of greater existential import—a knowledge of *motivations*.

It seems, though, that the rubrics of "pictorialization" and "fictionalization" have not yet exhausted the presence of the past in image. For a historical drama may not yet be a memorial, but just a story, however informative or capable of evoking this past. In this regard the breakthrough of a perceptual phantasy needs to be taken a step further as we ponder whether there can be such a thing as a perceptual (and public) recollection: a *historicizing* experience that goes beyond pictorialization and fictionalization.

§3. Perceptual Recollection?

There are two aspects of theatrical representation that will serve as a starting point in discussing some forms of perceptual, public recollection. To begin with, evidently there is no theatrical representation without inter-subjectivity, not only because the theater generally speaking entails a plurality of subjects playing different roles—at the very least as actor(s) and observer(s)—but above all because, as an ideal formation, the drama exists inter-subjectively *for* a community of subjects.[33] Secondly, it is clear that this ideal formation as we know it is de-

---

(Aristotle, *Poet.* 1448b11–17, trans. I. Bywater. In *The Basic Works of Aristotle*, ed. R. McKeon (New York: Random, 1941), 1457.

33 "[The content of the artwork] is a formation produced by phantasy and is something phantasied with the same mode of appearance, and it is abidingly posited as that in stable acceptance—as figment, as object: In this self-identity it is actually something beautiful. What is identified by me, posited by me as an abidingly accepted object, can also be posited as an object intersubjectively: The ideally identical figment as an object is then an intersubjective object, something existing ideally and intersubjectively, which we can all claim as our own through the really objective being of the work in its physical embodi-

pendent on the transmission of its own tradition. Without a tradition and a community that participates in, upholds, and passes on the art of dramatic representation, drama cannot survive, not even on a conceptual basis.[34] Theatrical representation is thus not just an enactment of a drama, but also, in a sense, its own re-enactment.

These two aspects about drama incidentally tell us about history and about the formation of a historical consciousness.[35] For history and historical time go beyond an individual lifetime and individual remembering, as they require the interlocking of different individuals, different lifetimes, and eventually different generations through time. History is not "subjective" but "communal," and pertains to communities changing through time. A community may be grasped as a collective of subjects belonging together in virtue of a common activity, unity of goals, and a shared commitment, thus contributing to and remaining in a tradition. A tradition, too, goes beyond the single subject – either coming into being *before* the subject or else remaining *after* him—even though it can neither remain nor be communicated independently of subjects. A tradition is inherited by a given generation, and handed down to the next generation, not just in gratitude for the past but also in concern for future generations. Some might say that the contents of a tradition are ultimately less important for a community than the fact that there is a tradition. And yet, the contents of a tradition are in no case unimportant, not only because the identity of a tradition cannot be a purely formal point, but also because a tradition includes its own domain of discourse and contextual clues which render it intelligible in practice and communicable across generations. Hence, questions of true and of falsity, of what has happened and of what is to be believed, pervade our historical concern, even if the answers to be given to these questions are not without personal and communal limitations. Orthodoxy and heterodoxy matter on the level of what is handed down, and not just in a religious tradition but also to a greater or lesser degree in all traditions. Again, an example of this is the fact that we have already mentioned, namely,

---

ment" (545). See also Éliane Escoubas, "*Billd, Fiktum* et Esprit de la Communauté chez Husserl" *Alter: Revue de Phénoménologie* 4 (1996) 281–300.

34 Cf. the short story by Jorge Luis Borges "La Busca de Averröes," a tale about the inevitable failure of Averröes to grasp the meaning of the terms "tragedy and comedy" in Aristotle's *Poetics* given the absence and prohibition of dramatic representation in Islam. (Jorge Luis Borges, "La Busca de Averröes." *Obras Completas* V. 1 [Bogotá: Planeta, 2007], 700–707).

35 In what follows I take my bearings (and some of my departures) from the second part of Rudolf Bernet's essay "My Time and the Time of the Other," In *Self-awareness, Temporality, and Alterity*, ed. Dan Zahavi (Dordrecht: Kluwer, 1998), 137–149.

that without the necessary contents or clues passed down by an empirical tradition of drama, its highly complex images would remain inaccessible.

From these very summary pointers with regards to history and tradition one can discern a likeness of history to a communal form of remembering. And yet, there is also a point of difference between history and collective personal remembrances. As Rudolf Bernet puts it,

> It is true that historical tradition, unlike individual remembering, appeals to other persons and to external aids such as the language of a story and of writing, or the presence of traces and monuments. It is also true that history as collective remembering is not merely a matter of conscious representation but rather of the interpretation of documents and ritual communal activities.[36]

History as collective remembering cannot be reduced to multiple, individual rememberings. Rather, collective remembering requires external mediation, and so, it is not surprising that images are found at this juncture. For images are also documents subject to (historical) interpretation, and moreover, images underlie ritual communal activities. When an image not just triggers subjective rememberings but also internally mediates the communal remembering of the past by giving a frame to past objects and experiences[37] the image may be called a *memorial*, while the individual acts corresponding to them may be called acts of *perceptual remembering*.

In order to grasp the common features of *memorials*, let us outline our proposal.

First of all, memorials are not reducible to mere pictorializations of a past event—even if pictures give rise to an awareness of "that has been" and in turn elicit historical interpretation *as* documents. We can give at least two reasons for this non-reducibility. On the one hand, a pictorialization such as a photograph does not *eo ipso* reanimate the past for a community or reconnect it to a communal present and future. The opposite in fact may be the case. As Roland Barthes says vis-à-vis the Winter Garden Photograph, "It exists only for me. For you, it would be nothing but an indifferent picture, one of the thousand manifestations of the 'ordinary.'"[38] On the other hand, mere pictorialization seems to elicit the least communal engagement and effort, notwithstanding the fact that a pictorial figment presupposes inter-subjectivity. Images may be public, but they do not necessarily require a distinctively communal response.

36 *Ibid.*, 144.

37 Regarding the "framing" of past experiences see also Rudolf Bernet, "Framing the past: Memory in Husserl, Proust and Barthes," In The Husserlian Foundations of Phenomenological Psychology (Duquesne University, Pittsburgh, 1993), 1–20.

38 Barthes, *Camera lucida*, 73.

A second distinction is that *memorials* are also not reducible to mere fictionalizations, or even to fictionalizations which have a past epoch or time as their theme. A Victorian drama, for example, may present a very vivid image of its time period, but an audience need not take such a historical drama as a memorial of that time. Even dramatizations about the past can be carried out for purposes that have no ostensible relation to a communal past—from entertainment and illustration or theoretical modeling, to historical scientific research.

Thirdly, *memorials* may and often do include pictorialization and fictionalization, but it is doubtful that these imaging functions are employed principally to *analogize* the past. A monument such as the Lion's Mound in Waterloo, on the spot where the Prince of Orange fell, and commemorating Wellington's victory, employs the pictorialization of a lion, but is not a memorial of it. Of course there are also manifold symbolic intentions and conventional elements in this monument. And yet, the monument is not a mere symbol for the past. One would rather say that this memorial is rather an abiding *site* for the past *in* the present, where a past event of communal significance can be recognized and celebrated.

Another example is a commemoration of the Gettysburg battle where two parties fictionally re-enact the clash of Confederate and Union armies. This commemoration employs fictionalization, and yet it arguably has less to do with giving an image of how the battle may have taken place and more with participating communally and emotionally *in* that past. As Samuel Johnson puts it, "That man is little to be envied, whose patriotism would not gain force upon the plain of Marathon, or whose piety would not grow warmer among the ruins of Iona."[39]

Fourthly, a memorial is not quite a repetition of the past event, even if a ritual communal activity lends itself to—in principle endless—repetition. The memorial enables a collective remembrance of a historical past, but does not confuse past, present, and future. A re-enactment of the Gettysburg battle neither multiplies its casualties nor necessarily flares up the war between states. In fact, such a memorial based on re-enactment may play a rather cathartic role.

However, one could speak vis-à-vis repetition of a certain divergence between secular and religious memorials. For a religious memorial invites a collective, perceptual remembering of divine transcendence in history, which as divine and as transcendent exceeds both past and present, remembrance and perception, while at the same time calling *for* both. This is the paradox: that these memorials not only serve as witness to the past for the present, but also

39 Samuel Johnson, *Journey to the Western Islands*. Quoted by C. S. Lewis in "Men Without Chests," In *The Abolition of Man* (New York: Harper Collins, 2001), 7.

reveal a continuity of presence between past, present, and future—ultimately between time and eternity. In Catholic Theology, for instance, the Eucharist is not only a memorial of the Last Supper and of the Passion of Christ, but also literally its original (re-) enactment. As St. Augustine tells us, "That bread, which you see on the altar, sanctified by the word of God, is the Body of Christ. That cup (or rather what the cup holds) sanctified by the word of God, is the Blood of Christ. Through these the Lord wished to commemorate for us His Body and Blood, which He poured out for the remission of sins."[40] Indeed, for this community and tradition, this particular past person and time is veritably present anew, even if He is neither purely perceived nor purely remembered. This memorial is a ritual communal activity that gives rise to a "once again" consciousness, but one in which the past is not remembered just by being imaged or framed. The past is remembered by being present itself but non-intuitively—veiled in mystery. For the believer, He who is remembered is Himself present, not as some thing plainly perceived or merely remembered, but as One who abides concealed *in* the perceived species through the consecration ritual. In this mystery of faith dogmatic theology teaches that a substantial change takes place *before* our eyes, without our seeing it—yet still along with our perceiving and remembering of Him.

Readily we admit that such a religious phenomena pushes the phenomenology of perceptual memory we have proposed to its break point. Yet this limitation is in keeping with the phenomena themselves, and avows to the importance of memorials for the life of a community. Moreover, the manifold ways in which the past appears in images is revealing of our nature. That we seek for the past in images, as we remember and seek to be remembered, is far from making a necessary concession to nostalgia. Rather, it evinces a longing for immortality at play in remembrance. It is a claim already made by Plato: "Every one of us, no matter what he does, is longing for the endless fame, the incomparable glory that is theirs, and the nobler he is, the greater his ambition, because he is in love with the eternal."[41]

40 Augustine, "Sermon 227." According to *Augustine in His Own Words*, ed. W. Harmless, S. J. (Washington: Catholic University of America, 2010), p. 153.

41 Plato, *Symp.* 208d–e, trans. M. Joyce, in *The Collected Dialogues of Plato*, ed. E. Hamilton and H. Cairns (Princeton: Princeton, 1989).

References

Aristotle. *The Basic Works of Aristotle*. Edited by Richard McKeon. New York: Random, 1941.

Augustine. *Augustine in His Own Words*. Edited and translated by W. Harmless, S. J. Washington: Catholic University of America, 2010.

— *Confessions*. Translated by F. J. Sheed. Indianapolis: Hackett, 1993.

Barthes, Roland. *Camera Lucida*. Translated by Richard Howard. New York: Hill and Wang, 1982.

Bernet, Rudolf. "Einleitung." In Edmund Husserl. *Texte zur Phänomenologie des inneren Zeitbewusstseins (1893-1917)*. Edited by Rudolf Bernet. Hamburg: Felix Meiner Verlag, 1985.

— "Framing the past: Memory in Husserl, Proust and Barthes." In *The Husserlian Foundations of Phenomenological Psychology*, 1-20. Pittsburgh: Duquesne University, 1993.

— "My Time and the Time of the Other," In *Self-awareness, Temporality, and Alterity*, edited by Dan Zahavi, 137–149. Dordrecht: Kluwer, 1998.

Borges, Jorge Luis. *Obras Completas* v. 2 (Buenos Aires: Emecé, 2005). English translation: Jorge Luis Borges. *Labyrinths: Selected Short Stories and Other Writings*, edited by D. A. Yates and J. E. Irby. New York: New Directions, 1962.

— *Obras Completas* v. 1. Bogotá: Planeta, 2007.

Brentano, Franz. *Grundzüge der Ästhetik*, edited by F. Mayer-Hillebrand. Bern: Francke, 1959.

Brough, John B. "Husserl on Memory." In *Edmund Husserl: Critical Assessments of Leading Philosophers*, edited by Rudolf Bernet, Donn Welton, and Gina Zavota, 325-346. London: Routledge, 2005.

— "Picturing Revisited: Picturing the Spiritual." In *The Truthful and The Good: Essays in Honor of Robert Sokolowski*, edited by John J. Drummond and James G. Hart, 47-62. Dordrecht: Kluwer, 1996.

Carreño, Javier. "On the Temporality of Images according to Husserl." *The New Yearbook for Phenomenology and Phenomenological Philosophy* 8 (2008): 73–92.

Deutsches Wörterbuch von Jacob und Wilhelm Grimm. 16 Bde. in 32 Teilbänden. Leipzig 1854-1961. Quellenverzeichnis Leipzig 1971. Online-Version : *Deutsches Wörterbuch von Jacob und Wilhelm Grimm.* Accessed July 17, 2013. http://woerterbuchnetz.de/DWB/?sigle=DWB&mode=Vernetzung&hitlist=&patternlist=&lemid=GA00011.

Escoubas, Éliane. "*Bild, Fiktum* et Esprit de la Communauté chez Husserl." *Alter: Revue de Phénoménologie* 4 (1996): 281–300.

Ferencz-Flatz, Christian. "Objects with a past: Husserl on 'ad-memorizing apperceptions'." *Continental Philosophy Review* 45 (2012): 171–188.

Husserl, Edmund. *Husserliana 19/1 und 19/2. Logische Untersuchungen. Zweiter Teil. Untersuchungen zur Phänomenologie und Theorie der Erkenntnis,* edited by Ursula Panzer. Den Haag: Martinus Nijhoff, 1984.

— *Husserliana 23. Phantasie, Bildbewusstsein, Erinnerung. Zur Phänomenologie der anschaulichen Vergegenwärtigung. Texte aus dem Nachlass: 1891–1925*, edited by Eduard Marbach. Den Haag: Martinus Nijhoff, 1980; English translation: *Collected Works XI.*

Phantasy, Image Consciousness, and Memory (1898–1925). Translated by J. B. Brough. Dordrecht: Springer, 2005.

— *Husserliana 13. Zur Phänomenologie der Intersubjektivität. Texte aus dem Nachlass. Erster Teil: 1905–1920*, edited by Iso Kern. Den Haag: Martinus Nijhoff, 1973. Husserl, E. *Husserliana 10. Zur Phänomenologie des inneren Zeitbewußtseins: 1893–1917*. Edited by Rudolf Boehm. Den Haag: Nijhoff, 1969. English translation: *Collected Works IV. On the Phenomenology of the Consciousness of Internal Time (1893–1917)*. Translated by John B. Brough. Dordrecht: Kluwer, 1991.

Jansen, Julia. "Phantasy's Systematic Place in Husserl's Work. On the Condition of Possibility for a Phenomenology of Experience." In *Edmund Husserl: Critical Assessments of Leading Philosophers*, edited by Rudolf Bernet, Donn Welton, and Gina Zavota, 221-243. London: Routledge, 2005.

Kortooms, Toine. *Phenomenology of Time. Edmund Husserl's Analysis of Time-Consciousness*. Dordrecht: Kluwer, 2002.

Lewis, Clive Staples. *The Abolition of Man*. New York: Harper Collins, 2001.

Plato. *Symposium*. Translated by M. Joyce. In *The Collected Dialogues of Plato*, edited by E. Hamilton and H. Cairns. Princeton: Princeton, 1989.

—

Javier Enrique Carreño
Franciscan University of Steubenville · Austrian Program
Kartause 1
3292 Gaming · Austria
javier.e.carreno@gmail.com

The Limits of Communication: A Phenomenological Account of Indication and Expression in Response to Derrida

Kevin Marren

Abstract

The aim of this paper is to clarify the treatment of communication in Husserl's phenomenology and, in consideration of certain of Derrida's criticisms, advance further explanations of communicative activity. Following its source material, the paper begins with an analysis of indication and expression, arguing that Derrida is, in fact, correct in interpreting Husserl as positing a mostly functional separation between the two concepts. Building from this point, however, the paper reasserts the phenomenological primacy of the subject as the origin from which all expression arises, while attacking the view that the absence any intersubjective appearance of 'thinking expressions' in communication is troublesome for phenomenology. This is to conclude, contrary to Derrida's criticism, that a description of communication is entirely reducible to a description of the expressive and interpretative activities of an ego-subject, and that, therefore, even if Derrida is correct in saying that the phenomenological account of communication withholds meaning from communicative artifacts, he has not adequately considered whether communication remains possible, underpinned by the ability of the subject to bring meaning with him into interpretation.

Keywords: Husserl, Derrida, Phenomenology, Communication, Meaning, Subject

Derrida begins *Speech and Phenomena* with an analysis of Husserl's distinction between the terms (a.) *indication* (*Anzeichen*) and (b.) *expression* (*Ausdruck*). My investigation into these terms begins with systematic considerations that progress towards (c.) critical commentary of Derrida's interpretation of Husserl's 'theory of signs.' In my preliminary analysis, I attempt to identify the reciprocal relationship between what one might describe as an 'intimating expression' and what Husserl calls 'thinking expression' (*gedanklichen Ausdruck*).[1] Both are correctly described as expression, but, as I will argue, only because of their co-ordinated *significational likeness* to one another.

Since neither function of expression, as my analysis attempts to demonstrate, is capable, by itself, of expressing in the fashion supposed by ordinary language, that is, neither *transfers* meaning, I analyze the complementary relationship between the two functions, determining how that relationship is underwritten by significational likeness and how that can provide everyday communi-

1 Notably, I have opted to use Findlay's translation of *gedanklichen Ausdruck* as thinking expression because he seems to have understood that, for Husserl, expressions in thought are equivalent to thinking, and his translation emphasizes this important point.

cative activity with expressed meanings. I then attempt to provide criticism of Derrida's claim that Husserl totally excludes meaningful signs (*beduetsame Zeichen*), i.e., expressions, from communication by showing that, even though meaningful expressions necessarily reside in the privacy of intentional consciousness, they often arise through interaction with communicative artifacts and, correlatively, guide communicative activity in the production of such artifacts.[2]

(a.) Systematic consideration of indication—An indicating sign is essentially an actively synthesized, complex whole that, through the intuitive grasping of a present, passively generated entity, fosters the belief in something that is not immediately present. "A thing is only properly an indication," Husserl asserts, "if and where it in fact serves to indicate something to some thinking being."[3] Thus, he divides indicating signs between their passive and active components such that, as the eidos of an indicating sign readily permits, one can more easily regard the sign's passive, physical body without having to consider its associative relationship. In reading a book, for instance, sometimes one allows his attention to wander, in which case, while staring blankly at a page, he regards the text as its mere visible concatenation and not yet as indicating anything. One can, however, take up passively generated things into active consciousness and recognize them, for example, as cultural artifacts that he habitually associates with a meaning.[4] In this way, those indications, which are, through the act of recognition, proper indications for the first time, now have the potential to *motivate* one's believing that an instance of some non-present thing, e.g. something that can be meant in a particular way, is (elsewhere), has been, or will be the case. Further, the more engrained these habits of association, i.e., the more practiced an individual is, the stronger the motivating force. When confronted with word-marks, reading is the default activity for a practiced reader. Only in states

---

2 I would like to thank Dr. Ralph Shain for helping me in my earliest research on this topic and Dr. Andrea Staiti whose comments about an earlier version of this paper prompted me to narrow my topic to more profoundly deal with meaning in communication.

3 Edmund Husserl, *Logical Investigations,* trans. John N. Findlay (New York: Humanities Press, 1970), 270; henceforth abbreviated as *LI*.

4 I have simplified my analysis on this point. Note: as Husserl explains in *Formal and Transcendental Logic*, §16, 50–51, an inattentive or empty consciousness (one which is not intuitively fulfilled) can be either articulated or not articulated. That is, even should I take a text to be an artifact of culture, presenting, for instance, some judgment-activity to be followed, I might understand only emptily that the text is the sign of a judgment without associatively connecting my understanding to any judgment in particular, i.e. my own judgment carried out in originary activity.

of great inattention are word-marks synthesized entirely passively for such a person, and in that case, the physical residue is not properly indicative.

The example above focuses on written language, but *anything can be an indication that gives to consciousness some non-present thing through association.* It would be inappropriate, therefore, to speak of indicating signs as though they logically entailed whatever they indicated to consciousness; rather, by definition, indicating signs can only become connected with what they indicate through one's previous experiences. Husserl elaborates upon this conclusion in the *Investigations*, first asserting, "[Indications] do not function in thought as logical grounds, but work through connections which previous actual demonstration, or blind learning on authority, has established among our convictions, whether as actual mental states or as dispositions for such" (*LI*, 272). He then shows that one recognizes logical relationships through the immediacy of one's insight such that their *content* is present immediately—sometimes implicitly though sometimes explicitly (and, as I will argue, always in respect to some meant noematic core). Since indication operates associatively in terms of previously developed convictions to recall only what is absent, logical premises, which would already contain the content present in a corresponding deductive conclusion, could not be said to indicate such a conclusion. One should not confuse, however, the content of a conclusion with the process that discloses it. The content dealt with in logical inference is *immediately* present (and therefore unfit for indication), but cannot be articulated without temporal mediation. That is, we should not think the temporally distended explicating processes which undergird all logical inference gives what is implicit and explicit in an absolute moment or in a pure, unmediated presence; rather, the explicit articulation serves to single out and rearrange content intended retentionally in consciousness that becomes given as a discrete whole through a persistent 'ray of meaning' extending across the entirety of the temporally distended explicating process. This will become clearer in my discussion of expression as this is precisely what is at stake when one considers the role of meaning in communication.

For now, one should note that the above discussion introduces Husserl's concept of habit, which he expands upon in §42 of *Formal and Transcendental Logic* and in §32 of the *Cartesian Meditations*. Husserl says in the former, "What the judger fashions now in active thinking, he possesses from now on as an abiding mental acquisition: the active acceptance changes for him into habitual acceptance."[5] And he reaffirms this in the *Meditations*, stating, "With every

5 Edmund Husserl, *Formal and Transcendental Logic*, trans. Dorion Cairns (The Hague: Martinus Nijhoff, 1969), §42 f., 104; henceforth abbreviated as *FTL*. Cf. on this subject also *FTL*, §16, 51–52.

act emanating from [the Ego] and having a *new* objective sense, [the Ego] acquires *a new abiding property*."[6]

The manner of 'interpreting' an indication can become fixed as an abiding property of the Ego. The active association of a discrete happening with another requires, no doubt, an original judgment of sorts. Taking smoke to indicate fire means that, at some point, the former followed the latter in such immediate succession and so proximally, that one could hardly avoid judging (assertorically, i.e., 'smoke is happening with fire') the two as belonging together. From then on, at the very least, smoke or fire will have the capacity to remind one of that original experience, and, quite aside from any final inductive inference, the two will be *felt* as having that moment in common such that having one presently could bring to mind the other. Likewise, this applies to learning how to use a linguistic system of indications. For example, one first learns color words when such are presented concomitantly with colored objects, but eventually the colored objects can be removed whilst one retains his understanding of the corresponding color words. Further, one can interpret unfamiliar words through their association to familiar words and, in these cases, what was once unfamiliar can become known through its correspondence with the words that are already fixed in habitually determined interpretations. This makes the *productive recombination* of the corresponding senses indicated through familiar words possible.

The reception of any communicative artifact (i.e., a text or speech act), then, can serve as an original basis for forming an habitual understanding of a word insofar as reading and hearing can produce original judgments about words in context. Each word-sign, in respect of its passive elements and no matter how familiar it is already, can acquire a new basis for its association with each novel instance of its presentation in intuition. Husserl never makes this claim explicit, but it does follow from the above analysis and is evident in one's everyday experiences. That is to say, because the habitual associations underpinning the recognition of indications as active-synthetic unities are grounded in an original positing act that depends on the spatiotemporal concomitance of an indicating signifier with its signified, each novel context wherein a word-sign is reiterated, insofar as the word-sign is now presented as spatiotemporally concomitant with new word-signs, would serve, under normal conditions, to augment the habitually held correspondence between any signifier and its signified through the corre-

6 Edmund Husserl, *Cartesian Medications*, trans. Dorion Cairns (Drodrecht: Kluwer, 1999), §32, 100–101; henceforth abbreviated as *CM*.

sponding judgment of their mutual belongingness to the novel context.[7] Further, according to Husserl, *any* positing or judging, once carried out, becomes an abiding property of the personal Ego.[8] Therefore, every new instance of a word-sign, insofar as it presents a novel basis for positing this mutual belongingness of the word-sign with its new context, shall also serve to present a novel basis for forming an abiding property. Of course, this will only occur roughly in proportion to the degree of attention that the receptive experience commands. To give a limit, in cases of extreme inattention or inactivity, or when overwhelmed, there will be little opportunity for the Ego to posit anything, and thus new habits will not follow. Not every experience changes drastically one's individual and habitual manners of understanding an indicating sign, but all attentive reading or listening does.

(b.) Systematic consideration of expression—Husserl presents an account of expression in the *Investigations* that is difficult to interpret, and, while he warns of equivocations pertaining to the term,[9] he never quite overcomes these in that work. For instance, Husserl says in §9 of the First Investigation, "One should not properly say (as one often does) that an *expression expresses its meaning*," but in §12, he comes close to contradicting himself by stating, "*What an expression expresses*… relates to the contents of [conscious] acts, and primarily to the meanings, which are often enough said to be 'expressed.'" Depending upon how he understands the inverted commas in §12, Husserl either rescinds his original position, or obscures the meaning of his analysis with equivocation. He admits succumbing to the latter, conceding later in §12, "The terms 'meaning,' 'con-

7 The signifier would become immediately associated with the new context and, insofar as it would also remain that signifier, would mediately introduce an association between what it previously signified and the new context. See Husserl's treatment of signifiers as signs of themselves in *LI*, Sixth Investigation, §14, 711.

8 Cf. *FTL*, 51. Husserl contends this point, but does not make the connection to the constant flux of associations that torment all indicative formations.

9 See *LI*, First Investigation, §5, 275, §8, 278–279, and §§12–14, 286–291, for Husserl's discussions of equivocation. Notice, however, that Husserl contradicts his own analysis. He says infamously is §8, which is problematic for many reasons, that in solitary mental life, expressions cannot function indicatively, and are, thus, not intimately joined to indication, but in §14, states that "Every expression intimates something…" Husserl is unclear in §8 about the essence of expression, and while one cannot ignore the section altogether, one can admit that he treats better elsewhere the issues that section raises. Besides, expression meant as the articulate explication of noemata, cannot be essentially one with expression meant as appearing in dialogue unless, somehow, the nomata could subsist in the physical, unthinking words themselves. Much to my dismay, §8 seems (though perhaps only seems) to argue that such is possible.

tent,' 'state of affairs,' and all similar terms harbor such powerful equivocations that our intention, even if expressed most carefully, still can promote misunderstanding." Doubtless, the term 'expression,' defined by Husserl as *meaningful signification*, will therefore harbor difficult to suspend, equivocal senses.

To combat the ambiguities inherent in our everyday-language understanding of the term 'expression,' an understanding that, despite his philosophical astuteness, still suffuses portions of Husserl's analysis, I intend to divide the term 'expression' between the two divergent functions it names:

(i.) *intimating expression* and (ii.) *thinking expression*. Thereafter, I intend to discuss expression more generally, considering whether the division of function constitutes, in this case, a division of essence or whether a reunification under one essence is possible.

(i.) Intimating expression—For provisional intelligibility, I suggest the phrase 'intimating expression' to describe any concatenation of signs produced with an author's communicative intent. Communication, further, shall be here defined as the *externalization*, or *manifestation*, of meaningful signification coupled with a subsequent interpretation and understanding of the significational product. Husserl himself takes to define expression, somewhat more broadly, as simply *meaningful signification*, but also describes a narrower, functionally separate, sub-species of expression in §7 of the First Investigation that serves to '*intimate* the thoughts and sense-giving inner experiences of a person's communicative intention' (*LI*, 276–278). Thus, I shall employ the phrase, 'intimating expression' in designation of manifest signs serving a communicative function, while I shall reserve the more general term, 'expression,' for the broader concept of meaningful signs that might (or might not) operate communicatively.

As communicative signs, intimating expressions purportedly serve to publicize the inner thought-content of an author for some receptive party, i.e., another thinking being. Further, according to Husserl, an intimating expression can serve this communicative function only through there being some *rough* "correlation among the corresponding physical and mental experiences of communicating persons" (*LI*, 277), which, one might contend, would also correspond with the receptive party's ability to correlate the manifest signification with an articulation of his own inner thought content. This double co-ordination—the correspondence of an author's thought content with his speech-form and the superimposition of the author's speech-form upon his interpreter's thought content—will be necessary if communication, in some sense, is to be upheld as a viable possibility. The task here is to describe adequately this correlation. Alternatively, one might attempt to contend that private thought-content could itself become manifest or immediately presented to another, as though appearing within the sign rather than through it, but this is obviously absurd. Meanings might be indicated in correspondence with intimating expressions, but meanings themselves would remain held immanently as expressed noematic objects, i.e., the meant as meant. All meanings are meanings for someone; nothing is meant without some thinking being meaning it. I will discuss this again in my response to Derrida.

As Husserl contends, then, intimating expressions in communicative speech function as indications,[10] and intimating expressions, when manifestly appearing, are just *indicating signs*, serving to provide thought-content with expressive, significational form. Husserl's analysis in *Formal and Transcendental Logic* helps to support this conclusion. He asserts:

> In the case of *reading* or *hearing*... we have, it is true, a sensuous unity and mutual congruity of the seen or heard *verbal signs* in their *sensuous configuration*; but in ordinary reading, *we by no means have, combined with that, an accompanying articulation of actual thinking, of thinking produced from the Ego, member by member in synthetic activity*. Rather, this course of thinking proper is only indicated (by the passively flowing synthesis of the sensuous verbal sounds) as a course of thinking *to be* performed (*FTL*, 50).[11]

In this quote, Husserl does not use the term 'expression,' but he doubtless means to describe the receptive activity arising in correspondence with an intimating expression. The 'sensuous verbal sounds' and all like manifest significations are instances of intimating expressions, i.e., indications. This leads one to consider again the First Investigation where Husserl asserts that statements, spoken or otherwise, merely 'give expression' to mental act-processes, and that, as discussed above, it would be inappropriate to say that an *expression actually expresses its meaning* (*LI*, 281). Intimating expressions, as mere indicating signs, really only *become* associated with meanings (and thus Husserl calls them meaningful) when giving expression to consciousness. In so doing, intimating expressions *become* thinking expressions.

(ii.) Thinking expression—Husserl introduces the phrase 'thinking expression' in §1 of the Sixth Investigation, stating, "The expressive role in speech lies... not in mere words, but in *expressive acts*: these create, for the correlated acts to be expressed by them, a new expressive material in which they can be given *thinking expression*" (*LI*, 675).[12] Intimating expressions, as discussed

---

10 Cf. *LI*, §7 and especially the first paragraph of §8.

11 I will soon diffuse the here apparent opposition between reading and hearing.

12 The phrase *gedanklichen Ausdruck* appears only in this section (in proximity to the phrase *gedankliche Intentionen)*, but this does not make the notion unimportant as it cor-

above thus stand in contrast to thinking expressions. The former are merely the manifest configuration of word-signs appearing in communicative artifacts. Often, however, these word-signs 'give expression' to consciousness, and taken up therein become thinking expressions, giving to the thinking activity a representative or *token* physical articulation that corresponds with the explicit thought content. The phrase 'thinking expression,' corresponds directly with the carrying out of the *course of thinking* prescribed by an intimating expression. That is, intimating expressions must not only suggest the meaning--content of thought, but also provide it with a significational form in common with real communicative artifacts appearing manifestly in the external world. Expressed more concisely, intimating expressions have the potential, as manifest indications, to associatively direct one toward some specific content for thought, but also, as material unities, the capacity to give articulation to the immanent parts of the very same thought content. To think, for example, that "something is red," not only recalls (through habit) redness in species, but articulates the specific instance of thinking the redness, betokening with signs and isolating its red components in relation to the discrete red object.

My interpretation of 'expression' becomes clearer through the consideration of, and slight break from, Husserl's treatment of the notion in *Ideas I*. In the *Investigations*, expression and the meant content are related incidentally through the desire of wanting to express oneself and particularly through the purposeful conjunction of the expression and its meaning through an act of consciousness. By contrast, in *Ideas*, the expressive stratum is said to incorporate into itself its meant content through the essential inclusion of the pre-expressive meaning-stratum. Husserl says:

> The expressive [act-]stratum [(*ausdrückende Schicht*)]… is perfectly identical in essence with the stratum undergoing the expression, and in the coincidence takes up its essence into itself to such an extent that we can call the expressive objectivating just objectivating itself, the <expressive> believing, <expressive> deeming likely, <expressive> doubting themselves, and as a whole, just believing, deeming likely, doubting, [etc.].[13]

This seems the entirely wrong way of thinking. That is, for example, in one's wishing act, one understands his wish all at once, such that the whole, pre-expressive wish-stratum becomes an implicit property subsumed by one's think-

---

responds with his earlier talk about expressions not expressing their meanings, but as being given to consciousness, i.e., as articulation of meanings. I have already mentioned this above and will give deeper consideration in what follows.

13 Edmund Husserl, *Ideas I*, trans. Fred Kersten (The Hague: Martinus Nijhoff, 1982), §124, 258.

ing Ego (at first actively, then habitually). Other impulsive mental events provide similar examples. While building a puzzle, for example, one need not verbally express to oneself the position of each puzzle piece, stating with each move (whether silently or out-loud), that "this piece fits thus and so," in order to have judged the piece's proper position. Likewise, one sometimes understands the meaning of what one wants to say before having expressed that meaning even to oneself. Thus, the subjective experience of the wishing and judging acts appear to consciousness differently from the expression of the very same wishing and judging acts. *Expressing something and meaning the same thing* are essentially separable acts precisely because the latter requires nothing of one's inner voice. The unarticulated, homogenous having of the wish or judgment constitutes the wishing or judging processes in its entire unarticulated duration. Thinking expressions, though, require an articulated duration, whereby a voice that mirrors the flowing of external discourse is given to consciousness.[14] Therein, the expressed meaning as meant remains as a persistent reference throughout the articulation of one's inner voice, while the inner voice changes and fades across the duration of the meaning-process it expresses.

Rojcewicz takes a similar position concerning thinking expressions. He says, "There is no pre-expressed thought. Prior to being expressed, thoughts are no more than gropings in a haze" and "For Husserl as for all phenomenologist, words are not merely an outer packaging of already constituted interior thoughts… on the contrary, we need words just in order to have definite thoughts."[15] Thus, for Rojcewicz, as for Husserl, expressions are a prerequisite for thinking in general, and the term 'thinking expression' is, in some sense, redundant insofar as *all thought is expressive*. I must disagree, however, with Rojcewicz's claim that "expressions *give birth* to their meanings"[16] (a theme that he plays upon throughout his essay) because, as I have attempted to show, meanings are there before and after their expression. One can feel the force of a meaning asserting itself before one even has the chance to express oneself about it. Thus, while Rojcewicz is partly right that there is a sense in which meanings are held within expressive acts, he misses that meaning-cores are there before expressive acts, as the mental constituents actually serving to *impregnate* the expressions with content. One's expressive thoughts, in order to be expressive, must first conjoin with already meaningful act-strata that are not yet them-

14 For those deaf from birth, I would assume, rather than voice, that images give expression to consciousness; but for the blind and deaf, I have no idea what expressive structures, if any, might be substituted.

15 Richard Rojcewicz, "Husserl versus Derrida," *Les Collectifs du Cirp (édition sépciale)* 1 (2010): 234.

16 *Ibid.*," 234.

selves articulated,[17] and this joining is always contingent upon one's desire for expression. Sometimes, however, meaningful experiences escape one's capacity to express oneself. And similarly, expressive acts sometimes turn out to be less meaningful than one intends. In these and similar cases, the meaning act is more profound than the expressing act, illustrating both the essential separation and the willed connection between the two.

Certainly, and despite Husserl's discussion in *Ideas*, the articulated expressive strata and the unarticulated inexpressive strata of consciousness cannot be identical in essence (that is, the wishing and expressive wishing are not the same), although portions of Husserl's intention are preserved in my analysis. However difficult it has been to describe, thought content and its expression are given concurrently and meaningfully insofar as the meaning expressed persists across the entire duration of the corresponding expressive act-process, semantically filling the expression through the coincidence. While an entirely satisfactory analogy is lacking, roughly illustrated, expressions are to inexpressive act-processes as moving pictures are to reels of film. A film reel is unspooled in a projector, the pictures change, the movie progresses, but there is only the one reel, and it remains unchanged throughout the showing of its content. Likewise, the verbally articulated segments of the meanings expressed in thoughts are given in cinematographic sequence,[18] while the corresponding ideal meaning remains noematically throughout the thought's articulation. Misconstrued, however, this analogy could undermine one's understanding of the issue. That is, thinking expressions, themselves, do not rely upon moving-pictures, or any intuitive fullness (the total depicturing of their content); rather, the important features of this analogy are the persistence of the film reel in comparison with the persistence of a meaning, and the tran-sience of the moving-frames, in comparison with the transient rise and fall of sounded or written articulations.

Understood correctly, what I have argued here is that thinking expressions alone, because of their concurrence and active translation of meaning into verbal form, constitute meaningful expressivity in a narrow sense. Expressive acts, as Husserl contends in one passage discussed above, provide speech-forms with meaning, and this turns out to be because these expressive acts are, in a more than incidental fashion, connected to the selfsame noematic meaning poles as

---

17 See again, *Ideas* §124, especially where Husserl discusses how sense-explication does not require 'the minimum of expression' (*Ideas,* §124, 257).

18 Hopefully this example does not too strongly recall Bergson. A comparison between Husserl and Bergson would be interesting, but such would greatly exceed the scope of this essay.

the unarticulated processes undergoing the expression. This will become clearer though a general comparison of intimating expressions and thinking expressions.

Returning to more general considerations, the term 'expression,' when applied broadly to the two otherwise divergent expressive functions, leads one to believe that what makes meanings public in communicative activity is an important, incessantly voluntary (but perhaps habituated), relationship between the production of intimating expressions, which recall an associated thought content, and the articulations of the inner voice constituting thinking expressions. Husserl is keen to identify this relationship, stating in *FTL*:

> The verbal sounds have their indications, which, in themselves, refer interdependently to one another and are built one on another. The sounds conjoin to make the unity of a word-formation, which, in turn consists of relatively self-contained formations. Each of these is bearer of a unity of indication; and the whole is a self-contained unity, which has noetically the phenomenological characteristic of self-containedness and, on the parallel side (noematically) the phenomenological characteristic that consists in the self-containedness of an indicated significational unity, built correspondingly out of indicated formations (*FTL*, 50).

What Husserl has identified in this passage is the reciprocal interconnectedness between thinking expressions and intimating expressions, underwritten by what I call their *significational likeness*—a concept which, in a way, clarifies some things about intimation. Both functions of expression, in fact, produce self-contained unities, but nothing prevents an intimating expression from becoming, *in a manner of speaking*, a thinking expression, i.e., insofar as the self-contained significational unity of an intimating expression can be taken up into consciousness, conditioning the production of the fantasied sonorous material which constitutes the inner voice. Likewise, this conditioning often directs one's activity to produce intimating expressions in the likeness of thinking expressions such that the sonorous articulations imagined of one's inner voice become externalized in manifest speaking. Furthermore, one manner of sensuous formation can become associated with another kind. Written marks, when read, can give voice to consciousness upon receipt. Correlatively, the inner voice often *leads* in the production of writing. Nothing, however, essentially makes sounds the default articulating device of thought; rather, thinking in likeness of speech is, for most, a merely habitual convenience. On the empirical-psychological side, which has nothing to do with the phenomenological consideration of inner thought, the convenience and privilege of the voice might arise from human psychological nature; nothing metaphysical, however, prescribes this privilege.

Both thinking expressions and intimating expressions, thereby, are said only to have been produced voluntarily in *significational likeness*, e.g., when significations actually appearing in communicative artifacts resemble the significations that make up an internal discourse, and this is an *essential* possibility of the material comprising the sign itself. Thus, the term 'expression,' when broadly considered, should simply denote this significational likeness, qua essence of sign, as well as the interplay between the indication of meaning by intimating verbal sounds and the articulation of thought content by one's inner expressive voice. Certainly, and that is further, all reception of intimating expressions will direct conscious regard such that, to have an intimating expression will simply count as having a corresponding thinking expression. After all, a sign is only indicative when it serves to indicate something to some thinking being, and, therefore, every intimating expression, as an instance of indication, presupposes incorporation into the life-stream of the thinking subject—nothing can be an indicating sign that is divorced entirely from this conscious activity. Thereby, every indicating sign that has likeness to one's inner expressive voice (or whatever articulative material one uses in place of one's voice) will be thought articulately in that likeness as the thinking expression it resembles, leading one simply to think expressively the content indicated by the intimating expression—such content as enters meaningfully into consciousness because of habitual association with the indicating sign.

Nevertheless, there are no meaningful expressions outside of consciousness and, further, no meaningful expressions manifesting in dialogue; rather, the intimating expressions in dialogue only become meaningfully *expressive* in, or, more properly, give expression to, consciousness, i.e., in consciously interpretative receipt of some communicative artifact. The further difference between intimating expressions and thinking expressions is that thinking expressions sometimes arise spontaneously as correlates to already meaningful act-processes in the immanent flowing of one's conscious life, and can guide one in the production of communicative artifacts, whereas intimating expressions necessarily *redirect* consciousness to recall associative connections in the production of new arrangements of old content. Still, the two functions belonging to signs share essentially the possibility of significational likeness; *they are, in fact, grounded in the essence of significational form.* This *common*ality between intimating expressions and thinking expressions provides for the possibility of limited *communication* that I will discuss through my commentary on Derrida.

(c.) Commentary on Derrida—Derrida presents a multifaceted critique of Husserl that seeks to overthrow the whole of phenomenology by disrupting its reliance on a certain theory of communication, and Derrida's criticism begins simply enough, offering an analysis of Husserl's so-called 'theory of signs.' Therein, Derrida contends that Husserl founds phenomenology upon an analysis

of signification that is fraught with crisis—an analysis that Husserl never approaches again after the *Investigations*.

Derrida builds his argument foremost in *Speech and Phenomena*, but provides a summary of what I believe is one of his most important claims in "Signature Event Context;" he says:

> The primary interest of the Husserlian analysis to which I am referring here (precisely by extracting it, up to a certain point, from its teleological and metaphysical context and horizon, an operation about which we must ask how and why it is always possible) is that it alleges, and it seems to me arrives at, a rigorous dissociation of the analysis of the sign or expression (*Ausdruck*), as a signifying sign, a sign meaning something (*bedeutsame Zeichen*), from all phenomena of communication.[19]

> The *crisis of meaning* (non-presence in general, absence as absence of the referent—of perception—or of meaning—of the actual intention to signify) is always linked to the essential possibility of writing; and this crisis is not an accident [...] but the positive possibility and "internal" structure of spoken language from a certain outside.[20]

In regard to indication and expression, this is certainly Derrida's most biting conclusion, but it is doubtful he ever proves that Husserl's identification of the functional separation between the two concepts pertaining to signification extends to overthrow the power of the significational likeness linking intimating expressions (indications) and thinking expression. Nonetheless, my previous analysis of the terms may appear to concede Derrida's conclusion through the admission that thinking expressions, i.e. meaningful signs, never appear outside of consciousness; and, I would be remiss if I did not explain myself further in this regard.

I will readily concede thinking expressions are not communicative in themselves because they presume an interiority to consciousness that cannot extend beyond the subject; but they are, at least, by virtue of their function, *always* meaningful. I contend that intimating expressions are never meaningful in themselves, but that they do always communicate insofar as they introduce, *through*

---

19 Jacques Derrida, "Signature Event Context," *Margins of Philosophy*, trans. Alan Bass (Chicago: Chicago Press), 319.

20 Derrida, "Signature Event Context," 319 (italics added). Derrida's notion of a crisis of meaning was already addressed where I contended that the privilege of spoken language is not metaphysical. I believe the reason I gave above suffices to answer some of the questions that might be raised here on the matter. Since I put writing and speaking on the same level the crisis, if it does not dissolve immediately, becomes generalized to the opposition between the 'internal structure of language' and its 'outside.' Thus, my analysis will more closely deal with this aspect, leaving Derrida's contention regarding the tension between writing and speaking for consideration in another paper.

association, meaningful content to consciousness. At first, this might seem identical to Derrida's conclusion that the "the pure function of expression is not to communicate, inform, or manifest, that is, to indicate,"[21] but, as I will argue, the pure function of expression, its essence, is really caught up in its significational form.

Derrida is arguing Husserl holds intimating expressions, being indications, bear entirely the burden of one's communicative effort, i.e., the effort which shows one's meaning to another person, whereas, *thinking expressions bear within themselves the entirety of one's meaning.* Therefore, according to Derrida, meaning evades communication by virtue of the functional separation within the essence of signification, and, aporetically, the manifest production of signs that makes artifacts communicative also renders such artifacts incapable of conveying the very meanings one was intending to relate (the meanings having both originated and remained in consciousness). The aporia, however, will depend on the assumption that, for communication to work, the locus of meaning must be the expressive product, i.e., the manifest signification, rather than, as Husserl contends, and as I have attempted to demonstrate, thinking consciousness as a meaning-stratum subsumed by expressive or interpretive act-processes. Thus, Derrida's entire criticisms rest upon the opposition between our everyday, naïve, treatment of *communication* as quite literally a transfer of meaning against the phenomenological treatment of communication as the correspondence and thoughtful expression of immanent meanings to manifest significational products.

Before one will accept that communication is not literally a transfer of meaning, though, one must understand how the phenomenological method demonstrates the primordial ownership of meaning in thinking expressions; and reading Husserl's later works (those after *Ideas I*) helps in illustrating this point. For example, Husserl provides a description of the phenomenologically reduced world in the *Cartesian Meditations*, illustrating how one's sense (that is, the meaning) of 'otherness' is really the sense of one's 'owness' in the mode 'over there.'[22] Likewise, thinking subjects constitute, within their respective spheres of owness, the meanings corresponding to an intimating expression, and each thinking subject constitutes what he takes to be another person's meaning in two steps mirroring the constitutional process giving otherness in general. One must first constitute the meaning of some communicative artifact from the original reawakening of a previous experience (thus coming to make it his own), and

21 Jacques Derrida, *Speech and Phenomena*, trans. David B. Allison (Evanston: Northwestern, 1973), 48; henceforth abbreviated as *SP*.

22 Cf. Husserl's Fifth *Cartesian Mediation*, esp. §§ 53–54, 145–149.

thereafter he must also constitute the original otherness of that experience's noematic meaning-content in recognition of something alien about its presentation (e.g., the involuntary redirecting of his thoughts). However, adequate description of this complex process is, doubtless, fraught with difficulties and blind corners because our everyday understanding, and thus, language itself, is built around the notion that meanings inhere in linguistic expression and that consciousness *finds* meanings within the words themselves. Phenomenologically, however, the reverse is demonstrated: consciousness provides meanings to language (and language provides one's already meaningful conscious affairs with articulation). Only when one correctly understands the subject as the original constituting basis for positing Objectivity, including the Objectivity peculiar to meanings, can one rightly begin to understand the phenomenological project and, through it, the limits of communication.

If one turns back to the *Investigations* for guidance on this matter, one will find Husserl asserting that the consciousness (really the personhood) of communicating persons becomes given through certain behavioral indications and that these indications, often enough, include linguistic signs that intimate expressivity. Husserl further states that a person, when interpreting, apperceives intimation in the same manner in which he apperceives the intimating person.[23] This entails, to use the terminology more common in the *Cartesian Mediations*, that another person's thinking expressions are *appresented* through some mediate, manifest entity—i.e., the intimating expression. I will return to this point momentarily. First, I would point out that Derrida also recognizes that this appresentation has a place in a phenomenological theory of communication, stating first in a summary of Husserl's position, and then in commentary:

> Outside the transcendental monadic sphere of what is my own (*mir eigenes*), the owness of my own (*Eigenheit*), my own self-presence, I only have relations of *analogical appresentation, of mediate and potential intentionality*, with the other's owness, with the self-presence of the other; [the other's] primordial presentation is closed to me (*SP*, 39).

> It now remains to be seen—and this is most important—in what respect expression itself implies, in its very structure, a non-plentitude. It is known as being more full than indication, since the appresentational detour is no longer necessary here, and since it can function as such in the alleged self-presence of solitary speech (*SP*, 71).

Thus, Derrida contends that appresentation deprives intimating expressions of meaning, whereas, in solitary mental life, where expression is had without appresentation, the thinking expressions remain filled. Without contesting Derri-

23 For instance, *LI*, 278.

da's reading, though, one should still come to understand better its limits. For Derrida, the 'appresentational detour,' a necessary aspect of communication, serves as the vanishing point of meaning in discourse, forever separating meant thought-content from the public domain. Thus, Derrida takes the logical conclusion of Husserl's theory of signs to contend that only our thinking expressions are meaningful and, *therefore*, that communication between persons, which converts thought into indications, is for phenomenology, an enterprise essentially detached from meanings. This conclusion, however, simply does not follow with the force Derrida intends. At its best, it is trivially true insofar as thinking expressions, meaningful expressions, occur only within or when taken up into the noetic processes constituting egological life. However, if what phenomenology demonstrates is that grasping another person's meaning-content first-hand was never, and could never really be the aim of communication, then Derrida's entire analysis, whether it correctly identifies that meanings cannot, for Husserl, appear directly in discourse, becomes an exercise in simply reading Husserl rather than offering full-blown criticism. That is, if communication was not supposed by Husserl as a means to make public what obviously cannot be made public, i.e., private thought content, and if people could understand one another even if thinking expressions became filled with meaning only in private life, then Derrida's criticism would fail to interrupt the pursuits of phenomenology. A multi-stage elaboration of this point follows.

In the *Cartesian Mediations*, Husserl explains that one encounters "animate organisms" in terms of their "changing but incessantly harmonious behavior" (*CM*, 144). This apperception of other ego-subjects is then verified as a certain otherness through the recognition of what Husserl calls an "originally inaccessible intention" (*CM*, 145) or, in other words, in there being something in the world peculiar to the existence of an ego-subject that the perceiving ego-subject does not recognize as having issued from, or belonging to, his sphere of owness (though that thing necessarily becomes something for him—an object appropriated for his own conscious regarded in the mode 'other'). The being in the world of this supposed thing confirms that it is not itself a meaning; meaning, as Husserl tells us, is identical to the content one thinks about in an act of expression, and precisely in the manner of a thought-content. That is, meaning, as confirmed by the First Investigation, is not its object referent, and not an object in the world, but, instead, a peculiar way of intentionally regarding an object referent.[24]

Certainly, then, in reception of a communication, one takes up intimating expressions that present an *alien arrangement* of significative formations and

24 Cf. on this subject, especially the *LI*, First Investigation, §§ 13–14, 289–291.

not an alien meaning. For example, when an interpreter reads another person's text, the word-marks guide the former's inner voice in articulation of an indicated content brought into consciousness through habitual associations and motivated by the sensuous formations of the text itself. That is, the manifest formations are also understood as indicative bodies, and content established in the interpreter's previous experience of such signs can return to his consciousness through their means; therein, the content becomes articulated following the flowing sensuous articulation of the intimating significations through which it was indicated. This propensity for the indicated noematic correlate corresponding to the reawakened experience to follow the linguistic articulations of the intimating expression is also an habitually developed predisposition usually based in one's having learned common language. At no point, however, are meanings either presented or appresented in communicative activity; rather, the noematic sense-cores arise in an originary fashion from one's habitually reawakened previous experiences, and then alterity becomes extrinsically attached to these cores *because the interpreter apperceptively recognizes the arrangement* of the intimating expression as unfamiliar or as having been issued from outside of his own originary activity. Thereby, all communicative acts serve to reveal other consciousness appresentivley, but not to appresent the *specific content* of the other consciousness.

In this way, meanings reside in one of two poles: either in the consciousness of an author or, in the consciousness of his interpreter; but never objectively accessible in something and the same thing for both. Derrida understands the modifying process of appresentation, but he misses the fullest implication of this understanding in his critique of Husserl. He famously asserts that, in French, meanings are given by what a 'sign wants to say,' and this *nonsense* phrase weakens his whole criticism. Strictly speaking, signs cannot *want* to say anything because they do not have desires or motivations. Only subjects *want* or *mean* to say anything and, correlatively, only they can avail themselves of the signified meaning through the spontaneous activity of one's inner voice or through the redirective activity of an interpretation. The sign itself is really nothing more than a passive entity that must be appropriated (as both a physical body and an indication of expressed meaning) through the conscious activity of an ego-subject. Thus, the sign says, first, exactly what its author intends it to mean—and in his voice—and then says what its interpreter takes it to mean—now in the interpreters voice—but neither has the other's meaning as constituted originally. Instead, both individuals must constitute the meanings appearing for them alongside their own thinking expressions that articulate the respective meanings and are informed by one's individual habits of association. The articulative material itself, although its arrangement appresents the other ego, is not the proper locus of meaning. Whether there is a continuity of expression from subject to subject is another question, however, and can only be addressed provided the owness of subjectivity can be defined and given limits.

The obvious limit, then, is one's learned behavior. Insofar as one experiences meanings as ideal unities, which, as Husserl asserts in *Formal and Transcendental Logic*, "after having, in fact, been generated, are still taken to be existent [such that] one does actually deal with them as with real physical things, even though they are far from being realities" (*FTL*, 71). Thus, one can develop habitual manners for associating the noematic content of his mind with linguistic formations; and he can, in turn, reproduce these formations at will and coupled with bodily acts of pointing in an attempt to teach others. One learns language in this fashion and one also teaches language in this fashion (and language is continually being learned and taught through the novel presentation of communicative artifacts as well). When one presents a child with a colored object, for instance, and speaks the corresponding color word, he is not communicating meaning content to the child, but rather a way of articulating that content which the teacher takes as originally presented to the child in the mode 'over there.' Likewise, when a competent language user engages in discourse, he does not *present* his interlocutor with meaning-content, nor does the interlocutor come to know whether or not the speaker means what he takes him to mean; rather, the interlocutor recalls his own learning of the speaker's words and constitutes such alongside the novel, other-appresenting, grammatical associations that they exhibit in order to frame his own expressive understanding of the other's meaning. What is other, though, is again, not the meaning, but the pure grammaticality of its significational articulation as given in the sensuous articulation of the intimating signs.

Nothing here is problematic or out of the ordinary, and Derrida's criticisms do not seem to undermine the phenomenological account of communicative activity in the slightest. It does not matter whether one individual understands his other in terms of a meaning, or whether pure expressivity is an internal happening that gives to consciousness only some new arrangement of previously experienced intuitional sense rather than an entirely alien content. What was incipient in previous sections of my essay is then brought into fullest view here: that is, the subjectivity of a person is primary and is the constituting basis for everything meaningful. Derrida understates this point, contending only that pure phenomenological subjectivity fails to comprehend what cannot be made present—namely the meaningfulness held by another person (and, in fact the otherness of the person, i.e. the noetic-noematic structure of his conscious life). Certainly, there is a limited sense in which this is correct, but nothing about it should ring as problematic for the phenomenologist. Communication, which is something

Derrida supposes the pure owness of the subject could not accommodate, is nevertheless a real possibility insofar as communication need only be what indication can provide, i.e., the articulative material for expressions, which then help form meaningful thinking expressions arising from habitually associated reawakenings of past experiences that become arranged according to these thoughts in terms of the novel articulative formations taken up from intimating expressions.

Further work in these areas is, of course highly in demand, and what I have attempted to show is modest at best. I have attempted to systematically reconstruct a reading of Husserl's so called 'theory of signs' through an original interpretation of Husserl's writings on the subject. I have also tried to demonstrate how Derrida's work, while a good reading of certain of Husserl's ideas, lacks merit as a criticism of the phenomenological account of communication. From this basis, and in refining my position as necessary, the next step is an analysis of communication in respect of its temporal character—another time, perhaps, in another paper.

References

Derrida, Jacques. "Signature Event Context." *Margins of Philosophy*. Translated by Alan Bass. Chicago: Chicago Press, 1982.

— *Speech and Phenomena*. Translated by David B. Allison. Evanston: Northwestern, 1973.

Husserl, Edmund. *Cartesian Medications*. Translated by Dorion Cairns. Drodrecht: Kluwer, 1999.

— *Formal and Transcendental Logic*. Translated by Dorion Cairns. The Hague: Martinus Nijhoff, 1969.

— *Ideas Pertaining to a Pure Phenomenology and to a Phenomenological Philosophy: First Book*. Translated by Fred Kersten. The Hague: Martinus Nijoff, 1982.

— *Logical Investigations*. Translated by John N. Findlay. New York: Humanities Press, 1970.

Richard Rojcewicz. "Husserl versus Derrida." *Les Collectifs du Cirp (édition sépciale)* 1 (2010): 233–248.

—

Kevin Marren
Philosophy Department
Boston College
140 Commonwealth Avenue
Chestnut Hill, MA 02467
USA
kevin.marren@bc.edu

Husserl on Type, Animal Life, and Cross-Species Morality

Sam Cocks

Abstract

Husserl's theories of type, association and analogical apperception serve to ground the basic form of community that arises between individual agents and other subjects, including other animals. The latter is because association and type are by nature inclusive, rather than exclusive, and do not function in the same manner as determinate concepts. Type is inclusive and indeterminate in the sense that it is not fixed, nor does it require exactitude. Rather it functions associatively in drawing out/together common resemblances and styles. I will argue that the cross-species relationships typification opens up contribute to the full constitution of the human monad, and that moral relationships with animals should follow the spirit of typification by being inclusive and open to indeterminacy. In this case, the term indeterminacy concerns moral decisions that do not require grounding in an exact understanding of what the animal may be thinking, desiring, etc. The essay will include a few remarks on the nature of analogy and description, as well as how my overall argument corresponds with similar deliberations in the study of animal cognition. All of the above concern interspecies and interpersonal interaction and how the formation of values and norms may occur.

Keywords: analogy, analogical apperception, animal consciousness, interspecies relationships, monad, morality, pairing, type, typification

1. Introduction

While it is true that our way of conceiving of, and relating to the natural environment has certainly matured, anthropocentric viewpoints have continued to blossom. For now, I will focus on our relationship with nonhuman animals. More specifically, while phenomenology and especially Husserl's variety of phenomenology aids us in approaching human and nonhuman relationships, Husserl is nonetheless a curious fellow in these regards. He made comments that set human and nonhuman animals at a stark divide, while also making several explicit and implicit remarks concerning certain subjective features humans and animals have in common, and how mutual understanding and human/non-human interactive relationships come about. My multistep thesis is rather straightforward. (1.) Husserl's theory of *type* and analogical apperception serve as the condition for the possibility of what are ultimately non-anthropocentric, but subjective centered, cross-species relationships. (2.) Becoming a human monad involves a constituted sense of self-identity and self-understanding—hence sense of personhood– that is often if not always, influenced through interaction with the nonhuman. As I will show, this includes an intergenerational component.

(3.) As Husserl's theory of type ties in with his theory of "analogical apperception," it is my argument that within those contexts whereby our imaginative modes of contemplation present us with reasons for extending moral concern to the animal, we should do so. This is because type and by way of extension, analogy, are in many ways *inclusive* and *indeterminate.* Inclusive in the sense of drawing the animal into an association with us because of some degree of meaningful similarity and indeterminate in the sense that type and analogy are by nature, not exact. Hence, whether or not our models of understanding are rigorously precise should not forestall moral consideration or action. Throughout the entire discussion I will keep Husserl's idea of the monad as illustrated in the fourth and fifth *Cartesian Meditations* in the forefront. All of the preceding will help us position Husserl as a thinker who presents a model of subjective life that involves a fundamental openness and receptivity to other species, the possibility of monadic enrichment through these encounters, the eventual presence of an intermonadic history, and the possibility of moral goodwill towards the nonhuman. It is important that the reader know that the current analysis focuses on Husserl's theory of type, our engagement of the non-human, roughly put, and how such ultimately concerns human experience. These same analyses could and should be applied to animal experience.[1]

1 I have a book length manuscript entitled "A Husserlian Phenomenology of Nature." There I go into much more detail about the implications of Husserl's thought for the natural environment, especially our relationship with non-human animal life. The book should surface sometime between years 2014–2015. One book and one article immediately come to mind as essential reading: Corinne Painter and Christian Lotz, eds., *Phenomenology and the Non-Human Animal*, (Netherlands: Springer, 2007), and Javier San Martin and Maria Luz Pintos Penaranda, "Animal Life and Phenomenology," in *The Reach of Reflection: Issues for Phenomenology's Second Century*, ed. Steven Crowell, Lester Embree, and Samuel J. Julian. (Center for Advanced Research in Phenomenology, 2011), electronically published at www.electronpress.com. As far as important texts that explore the full range of intersubjectivity see Dan Zahavi, *Self-Awareness and Alterity. A Phenomenological Investigation* (Evanston, Il: Northwestern University Press, 1999); *Husserl and Transcendental Subjectivity: A Response to the Linguistic-Pragmatic Critique*, trans. Elizabeth A. Behnke (Athens: Ohio University Press, 2001), and *Subjectivity and Selfhood: Investigating the First-Person Perspective* (London: MIT Press, 2005). Husserl texts that offer a considerable quantity of commentary on animals include: *Husserliana 13. Zur Phänomenologie der Intersubjektivität. Texte aus dem Nachlass. Erster Teil: 1905–1920*, ed. Iso Kern (Den Hagg: Martinus Nijhoff, 1973), *Husserliana 14. Zur Phänomenologie der Intersubjektivität. Text aus dem Nachlass. Zweiter Teil: 1921–1928*, ed. Iso Kern (Den Hagg: Martinus Nijhoff, 1973), *Husserliana 15. Zur Phänomenologie der Intersubjektivität. Text aus dem Nachlass. Dritter Teil: 1929–1935*, ed. Iso Kern (Den Hagg: Martinus Nijhoff, 1973), and *Späte Texte über Zeitkonstitution* (1929–1934): *Die C-Manuscripts*, Band VIII (Dordrecht: Springer, 2006).

2. Type and Pairing

First, a few remarks on Husserl's theory of monad. Husserl considers the *human* monad to be a person. This term designates concrete transcendental or subjective life. As a whole, a monad comprises both the subject as a history of position takings, and the world as lived and known within and through this history of position takings.[2] The monad is fundamentally underdetermined in the sense that interaction with other monads is crucial for the monad's full development. The fully real human person requires both incarnation and intersubjectivity. The other monads have a perspective of us that we ourselves cannot generate, are able to confirm or disconfirm what the world originally means to us as singular subjects, and are capable of expanding upon the sort of significance the world may possess. The latter includes us as persons, for the person is considered a member of the world and not only the transcendental condition for the possibility of a world. I will return these comments below. Now I will turn to type or typification for the latter is one of the fundamental mechanisms associated with the production of the monad.

Overall, type and typification concern the experience of recognizing similarities, and belong to the intentional structure of conscious life. Based on a previous experience of something, it is possible to experience future phenomena, in an associative manner, *as* similar in type. Husserl, not surprisingly, manages to utilize this idea in a variety of ways. As I will go into more detail in the next section, the leading type we will discuss is that of our own lived-body (*CM*, 111). However, something as simple as the pre-predicative, intuitive experience of what we come to call a dog can establish a type that then leads us to recognize other instances of dog-like things as similar. There is an instance of habituality at work here:

> ...the fact that such establishment of habitualities at every stage of explication, of coming to know an object in its particularities, is not something which concerns only this object itself; it is also that by which, at the same time, is prescribed a *type*, on the basis of which, by apperceptive transference, other objects of a similar kind also

2 Edmund Husserl, *Cartesian Meditations,* trans. Dorion Cairns (The Hague: Martinus Nijhoff, 1973), 67–68; henceforth abbreviated as *CM*: "From the Ego as identical pole, and as substrate of habitualities, we distinguish *the ego taken in full concreteness* – in that we take, in addition, that without which the Ego cannot after all be concrete. (The ego, taken in full concreteness, we propose to call by the Leibnizian name: monad.) The Ego can be concrete only in the flowing multiformity of his intentional life, along with the objects meant—and in some cases constituted as existent for him—in that life."

> appear from the first in a preliminary familiarity and are anticipated according to horizon.[3]

On type, Steinbock writes, "Typification is the process by which I render typical that which is unknown according to a certain *style*. Like optimalization, it is selective and exclusive. Even though we have not yet experienced diverse aspects of a thing in its particularity, typification prefigures similar concordant features, making precisely those (and not other) aspects that are foreshadowed particularly prominent, affective as expected."[4] Thus, type not only discloses the experience of one thing as similar to another, but also *emphasizes* the disclosure of what is similar. In this sense, type allows for a sort of experiential efficiency. However, while type *is* "selective" and "exclusive," it is also *indeterminate* in the sense that the way a previous experience and novel impression can serve to found a similarity is, even if somewhat rule bound, rather open. What falls into association by way of the type does not require the sort of specificity demanded by a strictly rule-bound concept.[5]

Husserl tells us that the type is "open indeterminateness, which is fulfilled only in the form of a *more precise determination*" (*EJ*, 125). When something similar is established, the type undergoes enrichment because of the novel contribution; there is a retroactive enhancement of the primary type. In any case, the enhanced type will both continue to function and expand throughout further novel experiences. On this, Lohmar notes, "we have found in the change of types, which is grounded in experience, one of the forms of the 'preservation' of pre-predicative experience."[6]

Finally, this activity is essentially pre-inferential, arises passively, is non-conceptual, and in the most original sense, concerns things given in concrete sensuous experience. The sense of similarity is not the result of a reasoned inference in that there is no conclusion that arises from thought out premises. Again—type functions in a passive and associative manner. As for its non-

---

3 Edmund Husserl, *Experience and Judgment: Investigations in a Genealogy of Logic*, trans. James S. Churchill and Karl Ameriks (Evanston: Chicago, 1973), 125; henceforth abbreviated as *EJ*.

4 Anthony J. Steinbock, *Home and Beyond: Generative Phenomenology after Husserl* (Evanston: Northwestern University Press, 1995), 159 (my emphasis).

5 Speaking of the subconceptual processing that is similar to typification, P. Gärdenfors states: "the patterns are not determined by a set of explicit rules as in symbolic systems but rather decided by various kinds of *approximations* and *optimizations*." Peter Gärdenfors, *Conceptual Spaces: The Geometry of Thought* (Cambridge: MIT Press, 2000), 239.

6 "Husserl's Type and Kant's Schemata: Systematic Reasons for Their Correlation or Identity," in *The New Husserl: A Critical Reader*, ed. Donn Welton (Bloomington: Indiana University Press, 2003), 113.

conceptual nature, one can see typification as performing the initial function of constraining experienced "information" so that the latter can be "handled on the conceptual or symbolic level." This is because original, perceptual experience involves material that is "too rich and too unstructured."[7] The latter is also pre-conceptual in the sense that it is not the product of an intersubjectively guided inquiry into the matters essential structure by way of eidetic analysis.

I will now turn to Husserl's theory of pairing in the fifth of the *Cartesian Meditations* to develop a more specific understanding of type. This will also allow us to return to the first set of remarks about monads. In this meditation, Husserl takes on the task of articulating how we come to know there is another transcendental subject before us. Our encounter with what we will come to call another subject begins when the other's physical body comes into prominence in my visual field, forming an associative relationship with my body (*CM*, 113). Following this, I transfer what I know of myself over to the other. That my body is "here" while her body is "there" maintains or stabilizes the difference necessary to posit unique subjectivities. [8]

What do I mean by the expression, "what I know of myself?" In the phenomenological sense, genetically speaking, the experience of otherness is dependent upon an individual's development. For instance, the individual subject's self-familiarity that is a component of pairing is historical in that it matures and becomes more complex through the individual's life. This also holds for the nonhuman. What I transfer over to the other, then, is highly dependent upon the latter self-understanding. Further, there are varieties of ways in which we can have intersubjective relationships that do not involve the mature sense of self-familiarity that Husserl is alluding to in the *Cartesian Meditations*.[9] The im portance of these for what I am considering cannot be overrated; however, such a discussion would take me too far afield. Nonetheless, as I will discuss in a moment, the degree of reflective and often concept-based self-understanding does not forestall the occurrence of associations between subjects that are based on pre-conceptual forms of self-comprehension. That is, the function of typification does not necessarily make use of predicative forms of comprehension, but

---

7 Gärdenfors, *Conceptual Spaces,* 221.

8 "...it awakens reproductively *another,* an immediately similar appearance included in the system constitutive of my animate organism as a body in space. It brings to mind the way my body would look 'If I were there'...the assimilative apperception becomes possible and established, by which the external body over there receives analogically from mine the sense, animate organism, and consequently the sense, organism belonging to another world," (*CM* 118) analogous to my primordial world.

9 See Gail Soffer, "The Other as Alter Ego: A Genetic Approach," *Husserl studies* 15 (1998): 151–166.

draws on pre-predicative and sometimes pre-reflective forms of self-understanding. Here I am thinking in particular of our pre-reflective awareness as embodied agents in the world. Thinking specifically of the claim of embodiment, one of Husserl's helpful contributions to phenomenology concerns his analysis of touching and being touched. Our bodies are not only open to a plethora of different experiences via sensation, but are also sensed as entities within the world. More, for Husserl and other standard phenomenologists, the latter includes a pre-conceptual awareness of subjective intentionality, including the cognitive and emotional dimensions associated with the latter. It is only through deliberative and intersubjective reflection upon subjective life that these dimensions come to fruition in all of their complexity.[10] The latter ideas are extremely important as concerns our interactions with non-human animals. I will come back to this in a moment.

While I am the "primarily institutive original" (*CM*, 112) in that I give or release to the other what I know of myself, it is crucial to remember an additional feature of the type discussed in Husserl's fifth meditation: the type itself is enhanced through the encounter with what is novel. I gain in self-understanding from the entity I serve to establish. Husserl writes:

> On more precise analysis we find essentially present here an intentional overreaching, coming about genetically (and by essential necessity) as soon as the data that undergo pairing have become prominent and simultaneously intended; we find, more particularly, a living mutual awakening and an overlaying of each with the sense of the other. This overlaying can bring a total or a partial coincidence, which in any particular instance has its degree, the limiting case being that of complete 'likeness' (*CM* 113).

A little bit later:

> Every overlapping at a distance, which occurs by virtue of associative pairing, is *at the same time a fusion* and therein, so far as incompatibilities do not interfere, an assimilation, and accommodation of the sense of the one member to that of the other (*CM*, 118).

There are manifold ways in which the subject is expanded within acts of associative pairing. For the time being I list three examples. (1.) In seeing another's *expressive* body, I come to know that my own incarnation is similar. I develop a more enhanced understanding of my external life, so to speak, in that my identity as an expressing, intentional agent becomes significantly more robust. (2.) I am recognized and secured as an entity *within* the world—I become a real per-

10 And typically, through a theory (i.e., the current discussion regarding HOT, or higher order theory).

son. (3.) I come to know the world possesses an objective presence that is not completely dependent on my own existence, and that verifies that what I do experience is not a perceptual illusion. One may object that how would I as primarily instituting type even be able to recognize the other as another subject if I did not already have some conception of myself as an object in the world? I remind the reader of my above comment on the touching-touched scenario in that the latter sets the stage for my initial encounter with the other subject. However, 1-3 appear to contribute to an increased understanding of that which made the experience possible in the first place. Interestingly enough, Husserl appears to think that the above forms of self-enrichment happen or can happen at the pre-conceptual level and are a result of associative functions. This doesn't seem strange given the fact that we find similar events occurring in a broad variety of animal lives.

3. Pairing and the Animal

How might this concern pairing based relationships between the nonhuman animal and ourselves? (1.) I remind the reader that type is indeterminate. The association that forms between two or more entities—here, lived-bodies—occurs through the presence of a common *style*. Essentially, similar behavior and overall bodily shape: "It is clear from the very beginning that only a similarity connecting, within my primordial sphere, that body over there with my body can serve as the motivational basis for the '*analogizing' apprehension* of that body as another animate organism" (*CM*, 111). The range of coincidence that can occur between the other and us is broad, and crosses between species. The significance of this is obvious—that which serves to bring others into a relationship with us—the type—is rather welcoming or inclusive.

> Higher psychic occurrences, diverse as they are and familiar as they have become, have furthermore their style of synthetic interconnexions and take their course in forms of their own, *which I can understand associatively on the basis of my empirical familiarity with the style of my own life, as exemplifying roughly differentiated typical forms.* In this sphere, moreover, ever successful understanding of what occurs in others has the effect of opening up new associations and new possibilities of understanding; and conversely, since every pairing association is reciprocal, every such understanding uncovers my own psychic life in its similarity and difference and, by bringing new features into prominence, makes it fruitful for new associations (*CM*, 120).

(2.) To remind the reader, the coinciding association I am currently speaking of is passive and automatic. The recognition of some range of similarity between the animal and us comes prior to rational arguments for and against how we may or may not have anything in common. Our awareness of the animal as familiar happens at a level *below* that of the relationship between things that possess an intersubjectively determined sense/concept.[11] (3.) In general, the structure of analogizing apprehension represents a kind of disarmed opening to the animal. Whatever it is we might say about the animal that could somehow construct more determinate barriers between us, comes after the irrefutable pairing. There is a primordial or relational function associated with our experiential life that undercuts all of our active manners to rationalize away our connection to the animal.[12] Yet on the other hand, and keeping the issue of indeterminacy in mind, the animal is not simply experienced *as if* they were human. To state that there is a sort of associational pull between animals and humans, in no way implies that they are somehow less exotic.

4. Interspecies Influence

Pairing with the animal is able to lead to an enrichment of the type as well as how it is we conceive of ourselves as monads. The pairing "enables me to learn from another truths which I only subsequently discover hold for me." Husserl is not "committed to a type of self-knowledge that is logically and temporally prior to others."[13] Thinking in regards to optimalization, pairing and hence typification is efficient in that it contributes to monadic development. I will discuss what I take to be four important examples that represent how our pairing association with animals contributes to the latter development. In addition, while one could argue the following examples are conceptually articulated, it is important

11 This most original way of relating to the animal inspires us to comprehend the latter in terms of some variety of lived-experience. See Corinne M. Painter on the importance of the animal being *unable* to think within the naturalistic attitude. The personalist attitude is primary for us, thus some variety must be for the animal. The naturalistic is a methodological abstraction. "Husserl and Stein: Empathy and Animals," in Painter and Lotz, eds., *Phenomenology and the Non-Human Animal*, 101–103.

12 Dan Zahavi's comment about Merleau-Ponty comes to mind, "we already find ourselves – prior to any adoption of a position – situated within an intersubjective world, and any struggle or conflict would be impossible and unthinkable if there were not already a common ground and a peaceful coexistence (*PP* 408/355; cf. *VI* 114–15/82–83, 298/244–245). Thus the other-as-object is merely an inauthentic modality of the other, and my objectification by the look of the other is only painful and unpleasant because it takes the place of a possible and more natural communication." *Husserl and Transcendental Intersubjectivity*, 154–155.

13 Frederick A. Elliston, "Husserl's Phenomenology of Empathy," in *Husserl: Expositions and Appraisals*, ed. Frederick A. Elliston, and Peter McCormick (London: University of Notre Dame Press, 1977), 224.

to remember that what founds these considerations is typification and associative pairing as discussed in the preceding section. (1.) At the minimum, through association with the nonhuman animal I gain a phenomenologically grounded, clearer understanding of myself as distinctively human *and* as part of a more inclusive continuum. There is the enrichment of the basic type "subjectivity" through an expansion of its parameters with each novel, cross species encounter. One could argue that this is a more phenomenologically grounded way of getting at evolution. Further, the recognition of belonging to a continuum can serve to unify the other epistemological and metaphysical claims researchers make about the varying manifestations of mind in the human and nonhuman animal: there is a range of mental life and the corresponding bodily expression. (2.) I develop a better understanding of my own behavioral practices through perceiving how the nonhuman lacks them. For instance, the human propensity towards self-deception that appears to be missing in the animal.[14] Whether or not humans should attempt to emulate this sort of animal behavior is a different question. (3.) The mutual functioning of type on both the side of the human subject and the animal legitimizes the presence of those sorts of interactions and forms of conversation that involve a sort of opacity. Type functions to open up the space for forms of communication that often possess a sense of subtlety and indeterminability. Of course, some forms of inter-species communication rather than others are often more replete with clarity. (4.) There are regularly enough, *type-produced openings* that allow for the appropriation or adoption of another's instantiation of a typical style. A particular style of individuality and with this individual way of relating to an environment is able to influence us. This would follow from the mutual transference that occurs within pairing (see above). We do not typically gain the latter through detached observation, but through engaged interaction. For example, spending time with animals in the many ways that humans do changes us. Sometimes this comes in the form of our own interpretation of what this animal offers me as a possible way of being a life. How we can be affected, and to what degree, has to do with the kind of resemblance between the animal and ourselves. Other examples that come to mind: what virtuous behaviors mean to us such as loyalty, patience, and unconditional affection, are enhanced and transformed through cross-species interaction. It would

14 Gärdenfors notes, "Apes, unlike humans, do not have any direct control over their movements or facial muscles. Those individuals among the homonids who could simulate their emotional expressions or suppress them were more often successful in deceiving others. There has consequently been evolutionary pressure to be able to control one's facial muscles and voice. This is a central part of what has been called the 'Machiavelllian intelligence.'" Peter Gärdenfors, "Slicing the theory of mind," *Danish Yearbook for Philosophy* 36 (2001): 11.

not strike me as odd to find that a researcher of wolves began to see matters as if a wolf.[15] To be sure, some of these examples eventually involve more of a sophisticated conceptual determination, but they are all built upon the space of association opened up through type, and the basic sorts of experiences found therein.

As a monad, 1–4 contribute powerfully to my own self-understanding and the possibilities I possess as a developing person. I realize in a broader sense what I am a part of (human as member of animal species), I realize my particular idiosyncrasies as a human, I expand to communicate outside of my own species, and come to inhabit or internalize the unique behaviors of the animal's I interact with.

These interactions lead to the production of a whole variety of postures and adjusted concepts, conditioning, overall, the way in which we intend meaning in the world. The latter includes sedimentation and intergenerational perpetuation. This perpetuation could be at the level of a concept, or, thinking more pre-predicatively and/or non-conceptually, at the level of the body and its expressions. Indeed this entire process is inherently temporal. Viewed slightly differently, a type foreshadows a *future* experience—in our case, the experience of the animal. However as the type has a backwards-forwards essence, what is eventually disclosed in the future ends up having an impact on our own self-understanding. This is then perpetuated again into the future, and ultimately, into the intergenerational future. The human monad or person, as an individual involving a fundamentally historical mode of self-understanding, is more than likely influenced by some degree of human-animal relationships. This may not necessarily be a relationship of immediacy; as stated, meaning undergoes varying forms of alteration and is then embedded within the cultural stream. Thus, personhood is dense twice over; it is the product of intergenerational influence, and this influence is in no way merely "human." None of this is to take away from what *is* distinctive about the human person in Husserl's eyes. Nevertheless, the function of type negates any easily demarcated border between the animal and us. Finally, that our humanity is influenced in some way through some kind of relationship to animals does not necessarily imply that an individual will have deep love for the animal. Indeed, this awareness could provoke even stronger resistance. I am currently drawing attention to the fact that what we understand as the human person, is in many ways permeated by a deep past of human-animal relationships.

As I hinted at earlier, the issue of the flexibility of type is important for animals also. I can only speculate, but would imagine that their interactions with us

15 Please note the "as if."

would or could continue to have repercussions with the others of their own community. There is no doubt that their typical experience is flexible. Otherwise, we would have to deal with the absurd question of how it is they are able to relate in any meaningful way to *us*.[16]

5. Interspecies Moral Orientation

Animal cognitive studies offer an astonishing array of models and tests that make it easier for us to grasp what the animal is experiencing. More specifically, concept based explorations of animal minds and behavior that take as their basis the latter pairing associations have led to the articulation of a large quantity of intersubjectively verified cross-species goods. I will take it for granted that some degree of mind is present in many animals, and that this fact is significant as concerns moral issues.[17] However, the specific differences in each species lived-bodily experience, the intergenerational, meaningful fabric of different species' home-worlds, and the presence of those animals with which we have a very limited relationship, draws a limit to the depth of our comprehension. First, and thinking of more theoretical forms of understanding, while I know my cat needs attention in order to remain biologically flourishing, and while certain people very familiar with cats verify that I am correctly interpreting particular emotional and lived-bodily clues, there is much to my cat's life that is, essentially, quite distant from me. Second, one can imagine a more complex moral context or situation involving thoroughly alien others that requires more imaginative and difficult forms of theoretical and practical deliberation. Third, and related to this last point, Dale Jamieson makes a crucial point about the importance of background knowledge and context in our comprehension of the animal:

16 "According to Husserl, animals only engage in the mere repetition of their specific environing-world, and not the generation or renewal of its meaning. If this were as far as a phenomenological scheme of constitution went we could not account fully for the constitution of intersubjectivity between humans and animals." Anthony Steinbock, "Generativity and the Scope of Generative Phenomenology," in *The New Husserl: A Critical Reader*, ed. Donn Welton (Bloomington: Indiana University Press, 2003), 311.

17 "Finally, attributing mental states to creatures is part of an outlook that recognizes them as morally significant. The relations between having a mind and being an object of moral concern are logically quite complex, but they are psychologically very strong. While favorable moral attitudes towards animals may not in themselves justify the ascription of mental states to animals, the existence of such feelings may lead us to see various facts about an animal as constituting or supporting such attributions." Dale Jamieson, "Science, Knowledge, and Animal Mind's," in *Proceedings of the Aristotelian Society, New Series,* Vol. 98, 1998, 99.

> Success in interpretation rests on many factors including background knowledge, appreciation of context, specific information about the creature in question, familiarity with his or her way of life, and general knowledge about the relationship between mental states and behavior. The rather bland fact is that knowing the mental states of others (whether human or animal) requires knowing what things are like around here.
>
> Mental attributions are based on behaviour but they occur against a large and complex set of empirical and conceptual structures.[18]

Fourth, no matter what studies might show, and no matter how far our rational analysis may take us, animals cannot speak to us.[19] Thinking of human-to-human interaction as an example, while we often believe we can easily read the bodily expressions of those whom we are closest, the fact is, we are sometimes wrong. It is my opinion that often, while not always, spoken and written language is crucial for successful human interaction and harmony.[20]

To elaborate further, and following Husserl, even when we remove layers or strata associated with human consciousness in order to develop models that might help us close in on a more precise understanding, we are still applying an analogy based on ourselves to something that is primarily similar and not the same. To be sure, our primary association with the animal is real and grounds our inquiries, but there are surely limits to our comprehension. Thinking back to what I mentioned earlier, I might come to see as if I were a wolf, but I will never see in a way that a wolf sees. To put matters more bluntly, while I am in no manner contesting the fact that we often have a good idea of in what way and of what the animal is conscious, we do not always have precise knowledge of the latter. In this section of the essay, the focus is on what might be *good* for the animal. That being said, what is interesting about analogy is the way in which it is similar to typification. To harken back to medieval philosophy, analogy, after all, does not carry the same weight as univocal explanation. Rather, it concerns resemblance, and thus, similarity. In my opinion, much as type is *inclusive* but *indeterminate,* I believe that more developed forms of analogy function in the same way.

More specifically, my argument is that moral decisions concerning animals should be inclusive in the sense that, even if we are unsure of exactly what might be appropriate for the animal, or whether or not we might fully grasp their intellectual orientation, they should be included in what Peter Singer noted as

---

18 Jamieson, "Science, Knowledge, and Animal Mind's," 92–93.

19 Of course, here one should consider the issue of sign language with primates.

20 Further, to assert that animals are less able to deceive in that their bodily expressions are 'more honest,' doesn't imply that it is that much easier for us to understand the full range of emotional content associated with the expression.

the "moral circle." Relatedly, the before mentioned inclusiveness should be based on the acceptance of indeterminacy. Moral inclusion does not necessarily require strict and exact conceptual determination or understanding. That is not to say that we should not try to develop the latter, but that it is not necessary in order to act. Finally, I believe that in so acting, we are acting in the spirit of the basic communalizing function we find present within typification.

On these last ideas, I remind the reader that pairing association is a form of automatic or instantaneous *inclusion*, with the inclusion involving, importantly, a strong element of difference. Not only the sort of difference associated with the co-presence of another mind that lacks originary givenness, but also, more importantly, the difference associated with another species. Further, what binds us to any specific animal is not the result of the application or projection of a rigid concept onto the other—this is not what we mean by inclusion - but rather a joining because of something that is common between us, a common style. Put otherwise, there is no subsumption of the entity under a univocal judgment or thesis. This is also true when we more actively apply a *conceptual* analogy to the other: we draw at least one thing into an inclusive relationship with another. Yet, the analogy functions more as a focus, as a way to see the phenomena, so to speak.[21]

Analogy as conceived in line with a phenomenological approach is also indeterminate and inexact. There are two ways this can be qualified. First, the models utilized within analogical comprehension are open, fluid, and flexible. As just stated, there is no rigid and determinate subsumption of an entity under a final and inflexible rule. Second, there is no expectation that the application of an analogy will result in what Husserl would understand as complete evidential fulfillment. This is not to say that analogical thinking is non-sense; analogies are based on an original similarity and we should indeed follow where these more explorative modes of thinking take us. Rather, the understanding simply lacks the kind of precision or exactness associated with what Husserl would more strictly call explanation. For Husserl analogy so conceived, is associated with description.[22] The latter does not pretend to obtain mathematical precision. Rather, following James Dodd, "a description… lacks stability by virtue of the fact that it proceeds more modestly than explanation, in fits and starts, cautious ad-

21 For instance, "God is like a lion." We utilize what we understand as a lion to approach that which we understand as God. As we make use of analogies to approach animals it seems as if the risk is much less.

22 Edmund Husserl, *Ideas pertaining to a Pure Phenomenology and to a Phenomenological Philosophy*, trans. Fred Kersten (Boston: Martinus Nijhoff Publishers, 1983), 138–139.

vances."[23] Still though, analogy as description "follows the given within its range,"[24] and in that manner, does not simply run off the tracks. What would be the range in these cases? What we believe we know about animals by way of association, theoretical reflection, and experiments.

Finally, in following the latter advice we embrace a parsimonious view. That is, instead of permitting the complicated limits surrounding our understanding of the animal to restrict our moral considerations, we embrace a more simplistic set of ideas or theories and the corresponding moral direction they generate. Simon Fitzpatrick is helpful here in distinguishing different senses of parsimony. We do not embrace "simplicity as (less) cognitive sophistication," but rather "simplicity as analogy," and "simplicity as evolutionary parsimony."[25] We reject the first option as it suggests fostering cognitive models that "attributes to the animal the 'lowest' or least sophisticated cognitive processes consistent with the available data."[26] Not only do the latter accounts often turn out to be too limited to account for the diversity of animal activity, but they also run the risk of permitting morally suspect behavior that goes against the grain of our recommendation for moral inclusiveness. Rather, I embrace the second and the third sorts of parsimony. The second notes that "the simplest explanation for similar behavior in different species is one that cites the operation of similar cognitive mechanisms," while the third "has argued that when closely related species such as chimpanzees and humans are found to behave in similar ways it is simpler, evolutionarily speaking, to posit *homologous* cognitive mechanisms in order to account for this similarity—shared mechanism inherited from a common ancestor."[27] To assert the presence of similar cognitive mechanisms and common ancestors does not then necessarily require us to assert that we know exactly what the animal thinks and feels. However, it does direct us to act in a way that is both morally cautious and morally generous. What might the latter mean?

In a general sense, this way of thinking requires us to offer the animal the benefit of the doubt regarding what we *do* know about the sorts of mind they might possess. This includes, among other things, the freedom to roam, think creatively, engage in social relationships, as well as cross-species interaction.

---

23 James Dodd, "Inner Life and Transcendental Philosophy in Husserl," *Tijdschrift voor filosofie* 66 (2004): 489.

24 Dodd, "Inner Life," 491.

25 Simon Fitzpatrick, "The primate mindreading controversy: a case study in simplicity and methodology in animal psychology," in *The Philosophy of Animal Minds,* ed. Robert W. Lurz (Cambridge: Cambridge University Press, 2009), 266.

26 Fitzpatrick, "The primate mindreading controversy," 266.

27 *Ibid.*, 267.

That a gorilla may appear calm in a zoo enclosure and even test "normal" in regards to certain kinds of hormone levels nonetheless belies what one *does* know of a gorilla's mind, and what they might require in order to live a flourishing life. The latter way of thinking will inevitably lead us to take more seriously whether we really *should* be eating them, arguments by J. Baird Callicott aside. Nonetheless, in so doing, our analogies function to include them in our moral considerations and do so based on an indeterminate sort of understanding.

6. Conclusion

The above endeavors to continue the discussion of Husserl's rich theory of type and how it concerns his theory of pairing. At the minimum, we saw there is an opening at the heart of experience for a relationship with the nonhuman that does not merely absorb this other into a delineated model. Further, the sorts of relationships that type enables serve the constitutional function of expanding personhood, while analogical and type-based thinking serve to motivate moral charity. The type induced intersubjective impact is as well, intergenerational. Ideas of concern I did not discuss here, or did not discuss in enough detail include: (1.) Husserl's rather elaborate theory of experiencing others in the sense that type and pairing require a certain amount of conscious sophistication. It would appear that there are modes of experiencing the other that do not necessarily function in the same way as the above. *How* this then alters the kind of relationships we have considered requires further discussion. (2.) A more detailed analysis of the way in which animals are altered and affected, and how this concerns their own kind of intergenerational perpetuation of meaning and subjective style.

References

Dodd, James. "Inner Life and Transcendental Philosophy in Husserl." *Tijdschrift voor filosofie* 66 (2004): 473–497.

Elliston, Frederick A. "Husserl's Phenomenology of Empathy." In *Husserl: Expositions and Appraisals*, edited by Frederick A. Elliston, and Peter McCormick, 213–231. London: University of Notre Dame Press, 1977.

Fitzpatrick, Simon. "The primate mindreading controversy: a case study in simplicity and methodology in animal psychology." In *The Philosophy of Animal Minds*, edited by Robert W. Lurz, 258–277. Cambridge: Cambridge University Press, 2009.

Gärdenfors, Peter. "Slicing the theory of mind." *Danish Yearbook for Philosophy* 36 (2001): 7–34.

— *Conceptual Spaces: The Geometry of Thought.* Cambridge: MIT Press, 2000.

Husserl, Edmund. *Cartesian Meditations.* Translated by Dorion Cairns. The Hague: Martinus Nijhoff, 1973.

— *Experience and Judgment*: *Investigations in a Genealogy of Logic*. Translated by James S. Churchill, and Karl Ameriks. Evanston: Chicago, 1973.

— *Husserliana 13. Zur Phänomenologie der Intersubjektivität. Texte aus dem Nachlass. Erster Teil: 1905–1920*, edited by Iso Kern. Den Hagg: Martinus Nijhoff, 1973.

— *Husserliana 14. Zur Phänomenologie der Intersubjektivität. Text aus dem Nachlass. Zweiter Teil: 1921*–1928, edited by Iso Kern. Den Hagg: Martinus Nijhoff, 1973.

— *Husserliana 15. Zur Phänomenologie der Intersubjektivität. Text aus dem Nachlass. Dritter Teil: 1929–1935*, edited by Iso Kern. Den Hagg: Martinus Nijhoff, 1973.

— *Ideas pertaining to a Pure Phenomenology and to a Phenomenological Philosophy*. Translated by Fred Kersten. Boston: Martinus Nijhoff Publishers, 1983.

— *Späte Texte über Zeitkonstitution* (1929–1934): *Die C-Manuscripts*. Band VIII. Dordrecht: Springer, 2006.

Jamieson, Dale. "Science, Knowledge, and Animal Mind's." In *Proceedings of the Aristotelian Society, New Series,* Vol. 98, 1998, 79–102.

Lohmar, Dieter. "Husserl's Type and Kant's Schemata: Systematic Reasons for Their Correlation or Identity." In *The New Husserl: A Critical Reader*, edited by Donn Welton, 93–124. Bloomington: Indiana University Press, 2003.

Painter, Corinne, and Christian Lotz, eds. *Phenomenology and the Non-Human Animal*. Netherlands: Springer, 2007.

San Martin, Javier, and Maria Luz Pintos Penaranda. "Animal Life and Phenomenology." In *The Reach of Reflection: Issues for Phenomenology's Second Century*, edited by Steven Crowell, Lester Embree, and Samuel J. Julian. Center for Advanced Research in Phenomenology, 2001. Electronically published at www.electronpress.com.

Soffer, Gail. "The Other as Alter Ego: A Genetic Approach." *Husserl studies* 15 (1998): 151–166.

Steinbock, Anthony J. *Home and Beyond: Generative Phenomenology after Husserl*. Evanston: Northwestern University Press, 1995.

— "Generativity and the Scope of Generative Phenomenology." In *The New Husserl: A Critical Reader*, edited by Donn Welton, 289–325. Bloomington: Indiana University Press, 2003.

Zahavi, Dan. *Husserl and Transcendental Subjectivity: A Response to the Linguistic-Pragmatic Critique*. Translated by Elizabeth A. Behnke. Athens: Ohio University Press, 2001.

— *Self-Awareness and Alterity. A Phenomenological Investigation.* Evanston, Il: Northwestern University Press, 1999.

— *Subjectivity and Selfhood: Investigating the First-Person Perspective.* London: MIT Press, 2005.

—

Sam Cocks
Department of Philosophy
University of Wisconsin-La Crosse
4110 Centennial Hall
1725 State Street
La Crosse, Wisconsin 54601
USA
scocks@uwlax.edu

Individuality and Incorporation: Renewing the Concepts of 'Life' and 'Biological Organism'

Elena Pagni

Abstract

Among the several questions that animate the current debate in Philosophy of Biology, one of the most important concerns the necessity to rethink of a series of 'common criteria' to explain 'life' and 'evolution of life.' As recently underlined by J. Dupré and M. A. O'Malley, 'life' seems to arise from a collaborative system of interactions between different forms of organisms. With regard to these considerations and the recent accounts of 'biological organism' outlined by Symbiogenesis, my primary purpose will be to reconsider, from a philosophical point of view, the conception of 'biological entity' as strictly dependent on complex levels of 'incorporation' among different beings.

Keywords: life, biological singularity, living organism, symbiogenesis, incorporation, intercorporeity

Introduction

As well stressed by John Dupré, biology should not be conceived as a science requiring "no more than the working of physics and chemistry;"[1] rather, he adds "there are emergent properties that could not have been predicted [...] from a complete knowledge of the constituents of a thing and their internal relations."[2]

Hereinafter, I give a brief account of some recent reviews of 'biological organism' pointed out by Theoretical Biology and evolutionary epistemology, thus presenting new philosophical issues for understanding *life phenomena* as singularities: by taking account of the role of 'symbiosis' and 'collaborations' between organisms through evolution, indeed, we see that the idea of 'biological entity' is strictly dependent on complex levels of incorporation among different beings.

In particular, following the Symbiogenetic accounts of evolution, many biological entities come out of a 'bodily fusion' among other organisms that live inside them as special guests or give rise to symbiotic associations. More-over,

1 John Dupré, "It is not possible to reduce Biological Explanations to Explanations in Chemistry and/or Physics," in *Contemporary Debates in Philosophy of Biology*, ed. Francisco J. Ayala and Robert Arp (Malden, Oxford, West Sussex: Blackwell Publishing Ltd., 2010), 33.

2 *Ibid.*, 33.

in the wake of what has been stated by Maurizio Casiraghi,[3] 'symbiosis' seemingly "deeply challenges the concept of individual," insomuch as Casiraghi may ask if "individuals are really evolving alone (i.e. is the real level on which selection acts?) or, rather, individuals are groups of living beings evolving as a single unit?"[4]

By this, in the final part of my paper I will focus on some issues of Symbiogenesis theory, pointing out the essential role played by 'symbiosis' in the emergence of new forms of life, therefore suggesting new philosophical perspectives in our understanding of 'living entities' and their co-constitution with the surrounding physical world.

1. "Extended criticality": symmetries breaking and randomness at the basis of the variability and singularity of living organisms

In *Mathematics and the Natural Sciences. The physical singularity of life*, Longo and Bailly claim:

> In biology, by way of contrast, one is struck by the enormous richness of structure with which living systems as given to us in phenomenal awareness are already endowed, while their theoretical formulation in terms of mathematical concepts suffices to model only certain aspects of that structure. And this is so in a manner which tends to fragment their organic unity and individuality and fails to do justice to their immersion in wider ecosystems. If we reflect on the role of mathematics in human cognition we are thereby led to re-examine its role in biology, since an analysis of living systems is the starting point of any reflection on cognition.[5]

In order to overcome reductionist biology and its tendency to reduce life phenomena to mere physical processes, the works of Longo and colleagues[6] point to new mathematical frames for analysing the biophysics of "life phenomena," in connection with their peculiar evolutionary history. In particular, as remarked by Longo and Bailly, in opposition to physical objects, "life phenomena" must be

3 Researcher in Zoology at University of Milano-Bicocca, Department Biotechnologies and Biosciences.

4 Maurizio Casiraghi, "Who are You? The Biological entities in the 21st Century Biology" (abstract of paper presented by Maurizio Casiraghi at the *3-Day International Conference on Evolutionary Patterns*, Lisbon, Portugal, May 27th – 29th, 2013).

5 Francis Bailly and Giuseppe Longo, *Mathematics and the Natural Sciences: The physical singularity of life* (London: Imperial College Press, 2010), 98.

6 Giuseppe Longo, Centre Cavaillès, CIRPHLES, CNRS – École Normale Supérieure, Paris; Francis Bailly (1939 – 2009, CNRS, Meudon; Maël Montévil (Dept. Cell and Developmental Biology, Tufts Univ. Boston).

seen as *structural singularities* that arise from a deep co-constitution with the surrounding environment.

As they note, "living matter" manifests "atypical" processes and dispositions that may not be completely reduced and explained by means of existing mathematical structures:

> [...] there are no sufficiently stable symmetries and corresponding invariants, as for phenotypes, which would allow to determine the evolutionary dynamics of the object. [...] the object follows a possible evolutionary trajectory, which may be considered generic.[7]

Thus, life phenomena appear characterized by permanent phases of transformation, associated to processes of proliferation with variability and permanent dynamics.

Moreover, while for physical objects, critical transitions represent "singularity" in a process, critical points of change in their global behaviour with respect to equilibrium (stasis), for biological entities, relative stability and coherence are preserved by means of *on-going symmetry breakings* and *extended critical transition phases*, through evolution.

Otherwise said, the biological dynamics related to organisms' development evolve along with permanent critical phases, characterized by permanent symmetry breakings and local/global mutual exchanges. Moreover, as Longo and Bailly clarify, biological matter may be compared to a "physical singularity "of long duration [...] the living element is in a progressive critical situation, a permanent passing between the local and global."[8]

With regards to this particular issue, Giuseppe Longo and Maël Montévil claim:

> Structural stability in biology, thus, should be understood more in terms of correlations of symmetries within an interval of the extended critical transition, rather than on their identical preservation. Yet, in biology, theoretical invariants are continually broken by these symmetry changes. A biological object (a cell, a multicellular organism, a species) continually changes symmetries, with respect to all control parameters, including time. Each mitosis is a symmetry change because the two new cells are not identical. This variability, under the mathematical form of symmetry breaking and constitution of new symmetries, is essential both for evolution and

7 Giuseppe Longo and Maël Montévil, "Extended criticality: phase spaces and enablement in biology," Preprint submitted to a Special Issue of *Chaos, Solitons & Fractals* (2013): 6.

8 Bailly and Longo, *Mathematics and the Natural Sciences*, 207.

> embryogenesis. The interval of criticality is then the "space of viability" or locus of the possible structural stability.[9]

On account of this, the equilibrium of living entities should be understood as dealing with permanent phases of *integration/regulation* and dynamic transitions between local and global different levels of organization. To clarify this point, Longo and Bailly claim:

> To summarize, integration is the presence (upwards-causally) of the local within the global structure, whereas regulation is the global structure causally affecting (downwards) local structures. It is this local/global interaction which organizes the "internal conduct" and which maintains the extended critical situation at least at the phenomenal level [...]. But besides having this role of stabilization, these interactions also participate in what we will later define as a (possible) theoretical indetermination of the evolution of living matter (both along phylogenesis and ontogenesis), because they interconnect different levels of organization [...].[10]

Emphasizing the relevant role played by Darwin's theory of evolution, by defining the *default state* of biological organisms as resulting from 1) "descent with modification" (due to ontogenetic and phylogenetic dynamics) and 2) "selection," Longo and Montévil note that, against Darwin, the reductionist approach of Modern Synthesis (*neo-Darwinism*) ultimately allows us to discard the role of "whole organism" in evolution: in MS accounts, in particular, we see that the pertinent observables are usually constituted by singular molecules and genetic components, without considering their interplay and integration with the organism *as a whole*.

> [...] the method of mathematical biology and biophysical modelling [...] is usually the transformation of a part of an organism (more generally, of a living system) into a physical system, in general separated from the organism and from the biological context it belongs to. This methodology often allows an understanding of some biological phenomena, from morphogenesis (phyllotaxis, formation of some organs...) to cellular networks and more [...]. [...] but it corresponds to a theoretical (and experimental) *in vitro* situation, and their relation with the cell is not understood by the physical approach alone. The understanding of the system in the cell requires an approach external to the structure of determination at play in the purely physical modelling. [...] these organs or nets, whose shape and dynamics are investigated by physical tools, are generally part of an organism. That is, they are regulated and in-

---

9 Giuseppe Longo and Maël Montévil, "From Physics to biology by extending criticality and symmetry breakings," *Progress in Biophysics and Molecular Biology* 106 (2011): 10.

10 Bailly and Longo, *Mathematics and the Natural Sciences*, 213.

tegrated in and by the organism and never develop like isolated or generic (completely defined by invariant rules) crystals or physical forms.[11]

As Longo and Montévil suggest, instead, since natural selection acts in correspondence of organisms in their environment, the actual pertinent observables should be considered organisms and phenotypes:

> We recall our choice of the biologically pertinent observables is based on the widely accepted fact that nothing makes sense in biology, if not analysed in terms of evolution. We summarize the observables as the "phenotype," that is, as the various (epistemic) components of an organism (organs, tissues, functions, internal and ecosystemic interactions…).
>
> Thus, evolution is both the result of random events at all levels of organization of life and of constraints that canalize it, in particular by excluding, by selection, incompatible paths—where selection is due both to the interaction with the ecosystem and the maintenance of a possibly renewed internal coherent structure of the organism, constructed through history.[12]

The key point of their analysis is that changes and variability, in biology, modify "the space of pertinent biological (evolutionary) observables, the phenotypes."[13] Furthermore, taking into account the increasing complexity of phenotypes during evolution (by sudden changes, transitions), Longo and Montévil introduce the notion of *enablement,* arguing that in opposition to the inert state of matter that is *inertia*, living beings interact with the surrounding environment "by acting (reproduction with modification and motility), use enabling conditions (are enabled by the environment), while producing new enabling conditions for further forms of life".[14]

Therefore, while in physics *variability* never concurs at the *intelligibility* and determination of the system, with *life phenomena*, *variability* is the key element for explaining *singularity* and *unpredictability* of life:

> […] biological dynamics are a complex blend of contingency (randomness), history and constraints. […] biological (constrained) randomness is essential to variability, thus to diversity, thus to life. […] In biology, randomness enhances variability and diversity."[15]

> […] the phenomenal life is characterized by the preservation of a few key invariants not only due to stability, but also to variability. For this reason, we have considered

11 Giuseppe Longo and Maël Montévil, *Biological time, Symmetries and singularities: Perspective on life* (forthcoming: October 2013), 14.

12 Longo and Montévil, "Extended criticality," 8.

13 *Ibid.*, 9.

14 Longo and Montévil, *Biological time, Symmetries and singularities*, 20.

15 Longo and Montévil, "Extended criticality," 7; 11.

> the intertwining and the interactions of organization levels on the one hand as one of the components of stabilization, and on the other as forming part of biological indetermination (as resonances destabilizing each level of organization). [...] Indeed, variability is at the centre of evolution and, in conjunction with individuation, it is also at the basis of ontogenesis. [...] On the one hand, a species presents a global structural stability, whereas its members may display different variations in their nature and behavior. Likewise, in its own time scale, an individual is (relatively) stable, while its cells evolve and die, each according to a different path. In one or the other case, the global structure is essentially (but not entirely) preserved while the local variations take place.[16]

Thus, as suggested above, the development of living beings as *singular forms of life* would result from permanent *criticality*, namely, from symmetry breakings, variability and unpredictability of evolutionary paths of ontogenesis and phylogenesis.

As stated, furthermore, while in classical and modern physics, mathematical symmetries are conceived as invariances with respect to a transformation[17] that allows fixing of "the space of the pertinent observables and parameters in which the theoretical determination of the systems takes place," with biological objects "no conceptual nor mathematical construction of a pre-given phase space is possible for phylogenetic trajectories"[18].

In connection with this, Longo and Montévil argue that the analysis of symmetry changes and criticality may be really relevant "for constructing a coherent biological knowledge"[19]: as a matter of fact, indeed, even if there is no possibility to grasp invariants for the intelligibility of variety and specificity of forms and functions of phenotypes, it would be possible to propose "general invariants" and "symmetries" for expressing the relative stability of biological rhythms, anti-entropy and extended criticality.

As Longo and Montévil remark, "anti-entropy will provide a possible quantification of phenotypic complexity and of its unbiased diffusion towards increasing values. It only makes sense in the presence of continual symmetry changes and enablement."[20]

On the whole, I consider Longo, Bailly and Montévil's insights into life phenomena relevant not only because they reintroduce the role of "organism" in the analysis of mechanisms of evolution, but also because of their understanding

16 Bailly and Longo, *Mathematics and the Natural Sciences*, 219; 220.

17 With respect to this particular topic, see Katherine Brading and Elena Castellani, "Symmetries and invariances in classical physics," in the *Handbook of the Philosophy of Physics*, ed. Jeremy Butterfield and John Earman (North-Holland, 2007), 1331–1367.

18 Longo and Montévil, "Extended criticality," 1–2.

19 *Ibid.*, 14.

20 Longo and Montévil, *Biological time, Symmetries and singularities*, 20.

of living objects in terms of *intercorporeal bodies*, namely, nothing more than "possibilities within spaces—ecosystems—in co-constitution," "historical, individuated, and not interchangeable;[21] ultimately, "the result of a unique history."

As I argue, the biological phenomena of 'extended criticality' and 'symmetries breakings' that—according to Longo and his collaborators—lie at the basis of life phenomena's development, may be well understood in the light of some recent reviews of 'biological entity' and 'life's evolution' advanced in contemporary biology, whereby 'life' appears characterized by permanent mechanisms of 'irreversibility', 'discontinuity' and increasingly bodily interactions between different forms of living beings.

2. 'Collaborative interactions' and 'symbioses': a new painting of 'biological organism'

(1) If we look more closely, some perspectives presented by Longo, Bailly and Montévil's works may be partially connected to certain discussions of contemporary Biology regarding the necessity to renew the status of "organism" in connection with the mechanisms of evolution.

In particular, as I argue, a new understanding of "biological organism" may throw new light on contemporary epistemological inquiry about the status of biological individuals with respect to their own peculiar evolutionary history.

As outlined by Huneman, in opposition to the relevant role that the concept of organism assumes in Darwin's evolutionary theory—"Darwin saw evolution as the transformation of the forms of organisms and this explanandum had to be explained by a natural selection that was targeting organisms"[22] in Modern Synthesis the role of organism has been completely omitted:

> Recent attempts to produce either an extended evolutionary theory or a novel kind of synthesis [...], indeed converge on the idea that "organism" was the neglected concept in evolutionary theory and that extending such a theory would provide an accurate understanding of this crucial phenomenon, as well as renew our general idea of evolution by putting organisms back in the center of the stage.[23]

As is well-known, in fact, beginning in the 1960s, neo-Darwinism has been focusing on the analysis of biological evolution making a synthesis of Darwinian

---

21 Giuseppe Longo and Maël Montévil, "The inert vs. the living state of matter: extended criticality, time geometry, anti-entropy—an overview," *Frontiers in Physiology* 39 (2012): 2.

22 Philippe Huneman, "Assessing the Prospects for a Return of Organisms in Evolutionary Biology," in *History of Philosophy of the Life Sciences* 32 (2010): 343.

23 *Ibid.*, 341–342.

natural selection with the progressing science of genes, their mechanisms of inheritance and mutations, and neglecting "the whole cycle life of an organism"[24] from the analysis of evolutionary processes.

In addition to it, Huneman adds, the dominant mechanistic path pursued up to now by Modern Synthesis, really allows us to conceive biological evolution in mere terms of "change of gene frequencies in populations,"[25] or, at least, to consider an organism a set of parts (populations of cells) characterized by specific causal relationships. In this scenario:

> The parts are somehow the descendants of previously autonomous living entities and the fact of organisms is basically the fact that at some point freely living cells entered into a process that yields multicellular life. Such a process is likely to be understood in evolutionary terms. So those biologists who are interested in explaining the biological fact of organisms as an historical result are saying that the kind of specific binding which ties the parts of organisms and which makes all of them somehow conditioned by all the others [...] results from evolutionary history, and can be explained in a neo-Darwinian manner.[26]

Even if—Huneman adds—an organism is essentially "multicellular," it cannot be reduced to mere multicellularity; effectively, an organism "includes many symbionts, at least, and most of them are needed for its development and its functioning."[27]

By this, it follows that there is a necessity to rethink of *biological organism* as resulting from "cooperative" and "heterogeneous" syntheses among different forms of life, emphasizing the relevant role of 'biological cooperation' and 'symbiotic associations' in the mechanisms of evolution:

> So even if the evolutionary transition program provides us with the best account of multicellularity, it does not give answer to the problem of the evolution of multicellular organisms. [...] If the evolutionary transitions program understands what is homogenous in organisms, namely multicellularity, those are also composed of heterogeneous syntheses, especially with all kinds of bacteria, which are not among the explananda of such programs [...].[28]

Likewise, against the reductionist principle that "the genome contains all information required to build an organism,"[29] John Dupré suggests understanding the

---

24 Steven Hawkins, *The Mechanisms of Evolution. A Critique of the Neo-Darwinian Modern Synthesis* (Reading, England: Light Press, 2011), 10.

25 Hawkins, *The Mechanisms of Evolution*, 344.

26 *Ibid.*, 362.

27 *Ibid.*, 365.

28 Hawkins, *The Mechanisms of Evolution*, 365.

29 Dupré, "Biological Explanations," 38.

DNA itself as a part of such exquisitely complex evolved system it interacts with:

> Strings of DNA, or even real genomes replete with histones and other structural elements that make up real physical chromosomes, do nothing on their own. Their involvement in the production of proteins is as part of a system that includes a very large number of additional molecules and cellular structures […]. It does certainly remain the case that the important activity of the genome is providing a template for transcription of RNA molecules, but which RNA molecules and in what quantities is dependent on interactions with many other constituents of the cell, and the actual structure of the DNA in the genome is constantly being modified by those other constituents. […] The capacities of DNA are not merely consequences of its molecular constitution, but are simultaneously determined by the systems of which DNA molecules are part.[30]

Moreover, according to David Morris, one may refer to the developing organism in terms of a "scratch pad" "on which the regulatory genome actually 'writes down' its current state": thus, a biological organism's growth should not be conceived as if only depending on a central machine, namely the regulatory genome, but as "the environmentally embedded self-transformation of one overall structure. This structure," Morris states, "comprises the regulatory genome *and* the organism's growing body."[31] Hence, Morris proposes defining the global structure of *regulatory genome* as a "self-articulating/self--transforming structure:"

> This structure encompasses both the regulatory genome and the growing body in its environment. […] the regulatory genome does not specify an already fixed, overall growth process, as if development merely reads out steps of a fixed plan. […] In this sense, development operates recursively, by transforming prior plans to develop *new* plans, where crucially, the plans in question are not abstract information, but actual bodies of organisms.[32]
>
> The developing organism is not a mere machine, with parts operating outside of parts […]. Nonetheless, in virtue of the regulatory genome's operation, this structure is internally, complexly, dynamically, and recursively articulate. This articulacy is key to developmental regulation, and exceeds any fixed set of finite states, since the articulacy in question can re-articulate itself, producing a new set of finite, or giving these states new, living meaning. Thus my concept of self--articulating/transforming structure, wherein the self-articulation of the structure can lead to its transformation

30 *Ibid.*, 38–39.

31 *Ibid.*, 99.

32 David Morris, "Merleau-Ponty, passivity and science. From structure, sense and expression, to life as phenomenal field, via the regulatory genome," *Chiasmi International* 14 (2013): 98.

> into a new articulatory structure, and self-transformation proceeds through new ways of articulating the system [...]. That is, what we have here is a system that we could never think of as autonomously constituting its sense, since sense production is always a labour that operates through inherited legacies that at the same time enable new sense to develop.[33]

In my view, the pivotal point of Morris's analysis concerns the challenge to reintroduce the role of "organism" as an essential and constitutive part of the mechanisms of evolution, by renewing the living body in terms of a complex structure on which the process of translation "is inseparable from the very growth of what it is that does the translating." In other words, the development of organism through evolution should be understood as the result of permanent and *collaborative* interchanges between inner (bodily) and external (environmental) dynamics.

Again, pointing to renew the common criteria of life, John Dupré and Maureen O'Malley highlight the fact of *collaboration* (between different forms of life) as one of the main important and functional features of *life*: in particular, they focus on the idea that the *metabolism* of living beings—say the whole set of biochemical reactions and transformations that is aimed to the self-sustainability of biological entities—is actually the result of collaborative affairs among multiple organisms, insomuch as *life* may be "found at collaborative intersections of many lineages."

> [...] metabolism is typically a collaborative activity involving many of things that are generally supposed to be discrete living entities. [...] It is hard to imagine life that is not collaborative, [...] both at the intracellular and the intercellular level, and we suggest that collaboration is, therefore, one of the central characteristic of life. [...] biological entities in our conception are series of dynamic and diverse collaborations, boundaries are flexible and unfixed.[34]

For Dupré and O'Malley, individual entities arise from a complex intertwining of "series of dynamic and diverse collaborations"—such as the environment and other forms of life (bacteria, viruses)—that play a fundamental role in genetic exchanges, multiplications and recombinations. The topic of *biological collaboration* opens new paths for distinguishing 'life' and 'non-life.' Indeed, when not collaborating, a biological organism may be considered "at most" as potentially alive:

> Life [...] occurs at the intersection of lineage formation and (typically collaborative) involvement in metabolism. Entities that are problem cases, such as viruses, can be

---

33 *Ibid.*, 99–100.

34 John Dupré and Maureen A. O'Malley, "Varieties of Living Things: Life at the Intersection of Lineage and Metabolism," *Philosophy and Theory in Biology* 1 (2009): 13.

> understood as alive when actively collaborating. When not collaborating, they have at most a potential life. [...] Countless non-cellular entities have individual life-histories, which they achieve through contributing to the lives and life-histories of the larger entities in which they collaborate, and this collaboration constitutes their claim to life.[35]

As one may see, the review of "life" in terms of *interdependent collaboration* leads us to reconsider the autonomy and singularity of entities as strictly depending on symbiotic collaborations rather than on mere self-organization. By this, Dupré and O'Malley identify 'life' with "a continuum of collaborativity."

> It is entirely reasonable to think of autonomy as centrally exhibited in collaboration rather than just rugged independence. Assuming that this kind of autonomy is what is needed to be a living thing, our account therefore includes viruses as not only living matter, but as full-blown living entities when they enter cells and interact with the cell's metabolic capacities.[36]

The concept of assuming 'life' in terms of what arises at the "intersection of lineage-forming, metabolically collaborative matter" and what is "organized within different interacting levels,"[37] recalls the Merleau-Pontian analysis of *intercorporeity*.

To illustrate this, for Merleau-Ponty one's own perceptual body (as perceptual flesh) should be considered a "prolongation and premonition" of another perceptual body. Ultimately, for Merleau-Ponty, the constitution of the sentient body already implies "an identity that is also a differentiation between what is my own and what is other."[38]

Maybe, this passage can be better understood by referring to what Scott Marratto states with regard to the Merleau-Pontian analysis of *intercorporeity*:

> [...] my sentient body, in its presence to me, as my own body, attests to what can never be present to me. It prepares me to 'experience' the other because it attests to that which can never be presented within any experience. The sense of my own embodiment is also a sense of my own non-self-identity and is thus a kind of opening to the presence of others. The my-own-ness, so to speak, of my body is enacted in the form of an expressive style that is reflective of my being-with--others and subtends my understanding of the behaviour of other.[39]

---

35 *Ibid.*, 13–14.

36 *Ibid.*, 15.

37 *Ibid.*, 16.

38 Scott L. Marratto, *The Intercorporeal Self. Merleau-Ponty on Subjectivity* (Albany: State University of New York Press, 2012), 145.

39 Marratto, *The Intercorporeal Self*, 142–143.

In addition to what has already been mentioned, we may also look at the biological phenomena of apoptosis and (controlled) cell proliferation and differentiation—by means of which the organism's life cycle is kept in equilibrium—to grasp that the whole development of biological individuals actually depends on essential "forms of collaboration" between different cellular populations. With particular regard to this topic, I consider the following statements of Philippe Huneman useful in grasping the complex relationship between the organism as a whole, its single parts and the whole set of *collaborative* processes involved in keeping the structure's stability during evolution:

> The recent biology, by discovering the process of programmed cell death, gives an unexpected echo [...]; at many levels, the life of an organism is given by the death of its cells. [...] the signs that produce cell death remind to the whole organism, which prescribes, in a way, according to its functional and/or developmental state, what cells must kill themselves.
>
> Thus, the death of parts is suspended in the life of whole, it determines this, but at the same time is commanded by this one. [...] Really, it's the same idea of biological individuality to be in play. For the reason that the relating categories here—outside, inside, whole, parts—are the same ones that allow speaking of living individuality. An animal, a plant, always determines an 'inside' and an 'outside;' and that arises by integrating several parts.[40]

Thus, an economy of cell death comes to be the condition of the individual life. By means of this, thus, by way of a process that only the correlative categories of individual or group allow to think of, the individual life becomes understandable by its relationship with the death of its parts.[41]

---

40 Philippe Huneman, "L'individualité biologique et la mort," *Philosophie* 102 (2009): 75 (my translation). The original text : "La biologie cellulaire récente, découvrant le suicide cellulaire, offre un écho inattendu [...]; à plusieurs niveaux, la vie d'un organisme se fait par la mort de ses cellules. [...] les signaux qui induisent le suicide cellulaire renvoient à la totalité de l'organisme, lequel prescrit en quelque sorte, selon son état fonctionnel et/ou développemental, quelles cellules doivent se tuer. La mort de parties est donc suspendue à la vie du tout, elle la détermine, mais en même temps se trouve commandée par celle-ci. [...] c'est en réalité l'idée même d'individualité biologique qui est en jeu. Car les catégories ici concernées – dehors, dedans, Tout, parties – sont celles-là mêmes qui permettent de parler d'individualité vivante. Un animal, une plante, cela délimite toujours un dedans et un dehors ; et cela se pose en intégrant en lui plusieurs parties."

41 Huneman, "L'individualité biologique et la mort," 77 (my translation). The original text : "Une économie de la mort cellulaire s'avère donc la condition de l'individu vivant. Par elle, donc par un processus que seules les catégories corrélatives d'individu et de groupe permettent de penser, la vie de l'individu devient compréhensible dans son rapport avec la mort de ses parties."

(2) As we know, in contrast to Modern Synthesis, Symbiogenesis theories underline the essential role of 'symbiosis' through evolution: in opposition to Modern Synthesis and the related idea that the principal vehicle of hereditary variation is given by random gene mutations; "the symbiogenetic theory [...] denies that natural selection is the dominant force in generating speciation and heritable variation;"[42] rather, Symbiogenesis looks at 'symbiosis' as the most relevant source of evolutionary novelty and mutations.

I will deal with some aspects of Symbiogenesis that I consider particularly relevant in view of acquainting new meanings of 'biological individuality' in relation to the history of an organism's evolution: in detail, I think of '*biological entities*' in terms of biological systems (combined structures) coming from bodily inter-collaborations between different beings; Symbiogenesis, indeed, occurs when the genome is factually 'embedded' and 'incorporated' by others.[43]

In *The Mechanisms of Evolution*, Steven Hawkins underlines the innovative insights of *biological individuality* ensuing from the 'symbiogenetic' theory outlined by Margulis and Sagan's analysis:

> [...] the major source of inherited variation is not random mutation. Rather, the important transmitted variation that leads to evolutionary novelty comes from the acquisition of genomes. Entire sets of genes, indeed whole organisms each with its own genome, are acquired and incorporated by others. The most common route of genome acquisition, furthermore, is by the process known as symbiogenesis.
>
> [...] Details abound that support the concept that all visible organisms, plants, animals, and fungi evolved by 'body fusion.' Fusion at the microscopic level led to genetic integration and formation of ever-more complex individuals.[44]

Hawkins concludes that, according to symbiogenetic theory it seems that the organism again takes possession of the role of primary unit of evolutionary selection.

As an alternative and in addition to Symbiogenesis, Hawkins also states the idea advanced by recent Developmental Systems theories that an organism's life-cycle should be assumed as resulting from interactions between multiple inheritance systems which are: a) genetic; b) non-genetic (i.e. internal to organisms); and c) dependent on the interaction between the organism and its surroundings.[45]

---

42 Hawkins, *The Mechanisms of Evolution*, 45.

43 As for the difference between 'symbiosis' and 'symbiogenesis,' we can say that 'symbiogenesis' usually occurs when long-term symbioses lead to a new kind of organism.

44 Lynn Margulis and Dorion Sagan, *Acquiring Genomes: A theory of the origins of species* (New York: Basic Books, 2002), 12; 56.

45 Hawkins, *The Mechanisms of Evolution*, 48–52.

> […] the life-cycle of an organism is constructed through joint-determination by multiple causes. Some of these causes are non-genetic but internal to the organism […]; others result from the interaction between the organism and its surroundings. […] Whilst the 'modern synthesis' asserts that environmental niches exist independently of organisms, and that random mutations in organisms cause them to 'fit' these niches, developmental systems theory asserts that organisms create their niches. The niche is created both by organism itself and by its ancestors.[46]

Hawkins claims, in contrast to Modern Synthesis, that "developmental systems theorists also make a strong case for taking the unit of selection to be the organism as a whole, and for abandoning the replicator-vehicle dichotomy. […] The claim," Hawkins continues, "that organisms are jointly determined by multiple causes in a process of self-organization looks set to become increasingly dominant…"[47]

Hence, Hawkins suggests a synthesis of Symbiogenetic and Developmental Systems Theory: namely, one overarching theory that explains the mechanism of evolution and the strict relationship between organisms, information (genes) and environment (niches) through evolution:

> In the Symbiogenesis theory natural selection is utilized in a minor capacity to explain the fine-grained adaptiveness of organisms to their environmental niches; the overwhelming majority of variation, and speciation, comes from symbiogenesis. In developmental systems theory there are no environmental niches that pre-date organisms, as organisms create their own niches.[48] By combining these two ideas we can postulate an adequate explanation for adaptiveness with no role for positive selection. It is symbiogenesis that generates variation and speciation, and the resulting organisms then create their own niches. We can then explain speciation and adaptiveness without positive natural selection. […]
>
> In this scenario there would be no organisms, no information, without an environment; but given that there is a species-created environment, there is a sense in which the genes that relate to that environment could be described as 'selfish'. This is because a species-created niche creates a high statistical probability that a given gene will be both created and lead to a given trait—this is a 'selfish' attempt to preserve the niche.[49]

In two very interesting articles, Nathalie Gontier highlights very clearly the strict relationship between 'individuality' and 'incorporation' (coming out from different forms of life cooperation) with respect to the evolution of organisms:

46 *Ibid.*, 52–53.

47 *Ibid.*, 61–62.

48 With regard to this topic see, for instance, Richard Lewontin, *The Triple Helix* (London: Harvard University Press, 2000).

49 Hawkins, *The Mechanisms of Evolution*, 62–63.

> Symbiogenesis is, as Margulis already pointed out "[i]ndividuality by incorporation," that is, a new distinctive entity emerges exactly because of the interaction. [...] Universal symbiogenesis is the process whereby new entities are introduced because of the interactions between (different) previously independently existing entities. These interactions encompass horizontal mergings and the new entities that emerge because of this are called symbionts. The process is irreversible and discontinuous.
>
> [...] symbiogenesis—which does not regard evolution as a sterile process—allows for cooperative views [...].[50]
>
> Symbiogenesis thus falls into place as a form of emergent evolution, known by the catchy phrase 'the whole is more than the sum of its parts.' Within universal symbiogenesis, different elements [...] that evolved along separate lines [...] somehow are combined and this new combination (a structure, element) has new properties that cannot be reduced to the parts that form the new structure.[51]

In conclusion, it seems to me that symbiogenesis theory succeeds in bringing together two important ideas concerning the mechanisms essentially involved in the biological evolution of organisms:

1. Every single biological organism is the product of a singular, historical and not interchangeable process;
2. Novelty and random mutations in life's evolution actually call for multiple types of intercorporeal and collaborative interactions between different entities.

Finally, as Evan Thompson expressively underlines in *Mind and Life*:

> [...] the single, individual organism as it is today is in a way an abstraction, both with respect to the organism as an ecologically embedded, developmental process, a life cycle, and with respect to the organism as a member of reproductive lineage. For this reason, we need to expand our framework to encompass development and evolution. [...] our living bodies, seen as a temporally extended lifeline, is a developmental process or life cycle that initiates new life cycles through reproduction. Living beings are constituted by historical networks of life cycles, and the units of evolution are developmental systems comprising organism and environment.[52]
>
> Nearly, every group of organisms has members that have formed symbiotic alliances with other organisms, including humans. The patterns of symbiont inheritance typi-

50 Nathalie Gontier, "Universal symbiogenesis: An alternative to universal selectionist accounts of evolution," *Symbiosis* 44 (2007): 175; 179.

51 Nathalie Gontier, "Evolutionary Epistemology and the origin and evolution of language: Taking symbiogenesis seriously," *Evolutionary Epistemology, Language and Culture* 39 (2006): 216.

52 Evan Thompson, *Mind in Life. Biology, Phenomenology, and the Sciences of Mind* (Cambridge: Harvard University Press, 2007), 166–167.

> cally involve the transmission from one generation to the next of whole functioning populations of symbionts, such as microbial communities, in addition to symbiont DNA (Margulis and Sagan 2002). These facts about symbiosis undermine the simplistic equation of biological identity through time with the transgenerational bridge of DNA.[53]

Thus, by assuming the idea that most '*biological organisms*' are 'embedded symbiotic structures' that arise from bodily collaborations and collaborative interchanges between different beings, how does our global understanding of physical nature change?

To simplify this, is it still possible to think of human beings as *individualities* or, rather, would it be better to conceive them in terms of 'symbiotic associations' of multiple beings that all together concur in manifesting one's singular existence?

Again, Symbiogenesis appears really to undermine the fundaments of reductionism and mechanistic theories that conceive living beings in terms of assembled machines:[54] rather, Symbiogenesis perceives combinations, mutations and fusions as the three main fundamental categories for describing organisms' life and their continuity as the result of a specific evolutionary history requiring increasing degrees of discontinuity (*extended criticality*). As Kozo-Polyansky previously noted:

> [...] they [combinations, mutations and fusions] assure not only leaps and gaps [missing links] but also continuity. Symbiogenesis, by fusion, obviously provides the largest scale of discontinuities.[55]

---

53 *Ibid.*, 177.

54 In particular I am referring to the mechanistic conceptions of biological organism arising from the analyses of Richard Dawkins, *The Selfish Gene* (Oxford: Oxford University Press, 1976), *The Extended Phenotype* (Oxford: Oxford University Press, 1982), and also Eric Davidson, *Genomic Regulatory Systems: Development and Evolution* (San Diego: Academic Press, 2001).

55 Boris Mikhaylovich Kozo-Polyansky, *Symbiogenesis. A new principle of evolution*, ed. by Victor Fet and Lynn Margulis (Cambridge: Harvard University Press, 2001), 113.

References

Bailly, Francis, and Giuseppe Longo. "Extended critical situations: the physical singularity of life phenomena" *Journal of Biological Systems* 16 (2008): 309–336.

— and Giuseppe Longo. *Francis Bailly and Giuseppe Longo, Mathematics and the Natural Sciences. The physical singularity of life.* London: Imperial College Press, 2010.

Brading, Katherine, and Elena Castellani. "Symmetries and invariances in classical physics." In *Handbook of the Philosophy of Physics*, edited by Jeremy Butterfield and John Earman, 1331–1367. North-Holland, 2007.

Dupré, John, and Maureen A. O'Malley. "Varieties of Living Things: Life at the Intersection of Lineage and Metabolism." *Philosophy and Theory in Biology* 1 (2009): 1–25

— *Human Nature and the Limits of Science.* Oxford: Clarendon Press, 2001.

Gontier, Nathalie. "Evolutionary Epistemology and the origin and evolution of language: Taking symbiogenesis seriously." *Evolutionary Epistemology, Language and Culture* 39 (2006): 195–226.

— "Universal symbiogenesis: An alternative to universal selectionist accounts of evolution." *Symbiosis* 44 (2007): 167–181.

Hawkins, Steven. *The Mechanisms of Evolution. A Critique of the Neo-Darwinian Modern Synthesis*. Reading, England: Light Press, 2011.

Huneman, Philippe. "Assessing the Prospects for a Return of Organisms in Evolutionary Biology." *History and Philosophy of the Life Sciences* 32 (2010): 341–372.

Longo, Giuseppe, and Maël Montévil. "Extended criticality: phase spaces and enablement in biology." Preprint submitted to a Special Issue of *Chaos, Solitons & Fractals* (2013).

— and Maël Montévil. "From physics to biology by extending criticality and symmetry breakings." *Progress in Biophysics and Molecular Biology* 106 (2011): 340–347.

— and Maël Montévil. "The inert vs. the living state of matter: extended criticality, time geometry, anti-entropy—an overview." *Frontiers in Physiology* 39 (2012).

Marratto, Scott L. *The Intercorporeal Self. Merleau-Ponty on Subjectivity*. Albany: State University of New York Press, 2012.

Morris, David. "Merleau-Ponty, passivity and science. From structure, sense and expression, to life as phenomenal field, via the regulatory genome." *Chiasmi International* 14 (2013): 89–111.

Thompson, Evan. *Mind in Life: Biology, Phenomenology, and the Sciences of Mind.* Cambridge: Harvard University Press, 2007.

—

Elena Pagni
Centre Cavaillès
École normale supérieure
29 rue d'Ulm
75005 paris
elenapagni@gmail.com

Dissatisfied Life: Philosophy of the Living in Canguilhem and Merleau-Ponty

Anton Vydra

Abstract

The paper concerns Georges Canguilhem's philosophy of the living being in dialogue with the ideas contained in Maurice Merleau-Ponty's phenomenology on the background of discussions about two heterogeneous camps in French philosophy in the period from the 1930s to 1980s. The author considers the works of both French philosophers in relation to certain biological issues. At first, he analyzes their approaches to problem of normal behaviour and the normal subject, and he shows that their notions of normality, even if very similar, are not completely identical. Next, he reflects on the concepts of "privileged behaviour" and "catastrophic reactions," which both philosophers borrowed from Kurt Goldstein. As a part of this reflection, the author addresses the problem of health and normality in a different way, one which leads to the concept of vitality elaborated by Canguilhem. The conclusions present an insight into the theme of dissatisfaction. The notion of dissatisfaction is present mostly in Canguilhem's writings, and it represents the subjective need of a person to harmonize the disrupted equilibrium between the person and his milieu, for example by means of knowledge. However, as the author shows in conclusion, Merleau-Ponty developed similar reflections on the role of questioning as a root of every philosophy and of research, by which a person harmonizes their own internal disorder.

Keywords: Canguilhem, Merleau-Ponty, normality, subjectivity

Discordances are here rather more meaningful and valuable.[1]

1. Introduction

In his well-known 1985 article, Michel Foucault distinguishes two types of French philosophy in the period from the 1930s to 1980s: Sartre and Merleau-

1 Georges Canguilhem, *Le normal et le pathologique* (Paris: Quadrige/PUF, 2011 [1966]), 17; *The Normal and the Pathological*, trans. Carolyn R. Fawcett (New York: Zone Books, 1991 [1989]), 46; henceforth abbreviated as *NP* with French/English page numbers. By this sentence we want to evoke the relations between the approaches of Canguilhem and Merleau-Ponty. In the original, it is used by Canguilhem within the different context of the distinctions between the approaches to physiology and pathology of August Comte and Claude Bernard on the one hand, and René Leriche on the other. In our case, I translate it differently from how it is in the English translation.
I would like to thank the anonymous reviewer for their critical and insightful comments about this article. These help me to prepare my contribution more coherently and to specify some of my assertions. They also offer me a good background to my considerations about Canguilhem and Merleau-Ponty.

Ponty on the one hand; and Cavaillès, Bachelard, Koyré, Canguilhem on the other.[2] The re-written version of the article appeared later as Foucault's "Introduction" to the English translation of *The Normal and the Pathological.*[3] Of course, the distinction between two types of philosophy is based on a strict opposition between philosophy of ego, subjectivity and experience and philosophy of knowledge and rationality. However, in both of the texts the main subject matter of Foucault's assertion is the reception of phenomenology in French philosophy. In the rewritten version Foucault even writes: "In other words, we are dealing with two modalities, according to which phenomenology was taken up in France [...]."[4] Thus, Foucault clearly declares the possibility of two different readings of the Husserlian project and consequently of two 'profoundly heterogeneous' philosophies. For the most part, it is possible to understand this reception of phenomenology in the case of Sartre, Merleau-Ponty, Cavaillès, Koyré and Bachelard, but it is not so clear in the case of Canguilhem. The stimuli for this doubt perhaps lie in his not very frequent usage of the notion of phenomenology, in the absence of any interpretation of Husserl's works, or in his rather severe words about Husserl as "an old, bitter man who has lost his impact in favour of Heidegger."[5]

However, the stimuli cannot be conceived as a sufficient demonstration that there is no phenomenological attitude in Canguilhem's writings. We can find a lot of extracts from his works in which the French philosopher considers sciences from a perspective in many respects very close to Husserl's *The Crisis of European Sciences and Transcendental Phenomenology*. For example, he emphasizes—really frequently—the role of the subjective experience and the meaning (*sens*) in sciences; he even points to it when he writes about the epistemologist's work in the history of sciences, in which he must 'reactivate the

2 Michel Foucault, "La vie: l'expérience et la science," *Revue de métaphysique et de morale* 90 (1985): 3–14. Robert Hurley's English translation was published as "Life: Experience and Science," in *Essential Works of Foucault 1954–1984 (Aesthetics, Method, and Epistemology)*, Volume II, ed. Paul Rabinow (New York: The New Press, 1998), 465–478.

3 Michel Foucault, "Introduction" to *The Normal and the Pathological*, by Georges Canguilhem, 7–24.

4 *Ibid.*, 8.

5 Georges Canguilhem, *Vie et mort de Jean Cavaillès* (Paris: Allia, 2004 [1976]), 16.

meaning.'[6] It is very close to some claims in *The Crisis*, especially to its famous *Beilage III* ("The Origin of Geometry").[7]

Nevertheless, the aim of my article does not lie in discussing whether or not Canguilhem is a latent phenomenologist; this is not important in order to do phenomenology. Rather I would like to focus on how Canguilhem's way of thinking could become fruitful for phenomenological research. Accordingly, Foucault's very strict differentiation of two camps in French philosophy does not seem to me so clear. For example, it is true that Canguilhem is not a philosopher of transcendental ego, but it seems to me false to say he is not a philosopher of subjectivity and experience.[8] We can indeed discuss two modalities of thinking about subjectivity and experience, but in relation to Canguilhem these two subject matters cannot play a role of '*differentia specifica*' by which we distinguish one modality of thinking from another one. We can emphasize different points of view, methods or approaches, but in this case it seems to be necessary to show how they function, especially if they are exposed to the same topic and the same interest.

That is the reason why I have chosen one representative from each camp: Merleau-Ponty on the one hand and Canguilhem on the other. Both of them consider very similar topics and problems in their writings on biological issues. For both of them biology was not only an inspiring science, but they also claim (in rare concordance) that contemporary biology tends to be a kind of philosophy [of life, as Canguilhem adds].[9] Both of them were influenced and fascinated by Kurt Goldstein's major work *Der Aufbau des Organismus*; and they developed

---

6 Georges Canguilhem, *Idéologie et rationalité dans l'histoire des sciences de la vie* (Paris: J. Vrin, 2009 [1977]), 25.

7 In relation to the role of subjectivity and the meaning in sciences, see also Derrida's consideration about Cavaillès who believed that the attitude of creative (and thus normative) subjectivity is impossible or "difficult to admit in phenomenology." According to Derrida this claim is not quite precise, because Husserl shows the subjectivity in sciences always simultaneously *normed* as well as *normative*. Jacques Derrida, "Introduction" to *L'origine de la géométrie*, by Edmund Husserl (Paris: PUF, 2010 [1962]), 158.

8 Accordingly, it would not be right to say that while Canguilhem belongs to the camp of philosophers of knowledge, Merleau-Ponty does not. I would like to refer to an interesting epistemological interpretation of Merleau-Ponty, in the conclusion of which the author claims that in reality Merleau-Ponty becomes a phenomenologist if he is understood epistemologically. Michal Lipták, "Raný Merleau-Ponty v diskusii s naturalizmom a psychológiou," *Filozofia* 66 (2011): 943.

9 Cf. Georges Canguilhem, "La nouvelle connaissance de la vie," in *Etudes d'histoire et de philosophie des sciences concernant les vivants et la vie* (Paris: J. Vrin, 2002 [1968]), 364; Maurice Merleau-Ponty, *Notes de cours 1959–1961* (Paris: Gallimard, 1996), 91 and also 388.

a lot of topics from this work in their philosophies, even if with quite different emphasis on what they saw as important. Merleau-Ponty as well as Canguilhem questioned the notions of organism, of animal—or more precisely—organismic behaviour, of the status of the human being in its relation to other living beings. Both of them stressed that to do biology today (i.e. a kind of philosophy), it is sometimes necessary to withdraw from the personal status of a human being and rather to feel like an animal. For example, in his *Phénoménologie de la perception*, Merleau-Ponty writes that "my organism, as a pre-personal cleaving to the general form of the world, as an anonymous and general existence, plays, beneath my personal life, the part of an *inborn complex.*"[10] What does this expression mean? Of course, the anonymity of an organism as a mere cell structure or a mass of proteins is obvious because such an organism has no singular name for itself (or maybe on its own). However, the organism is—as Colin Smith understands Merleau-Ponty's words—"a repository of acquired experience and habitual action."[11] The number of our own bodily reactions is not personal. They have a different origin, one in the organism itself. Our personal freedom can control it, eliminate it, manage it, but as *a complex* the organism remains an ineffaceable part of our existence. Hence, every discussion of our own human existence must also consider our organic life and bodily behaviour.

If we turn to Canguilhem we will find another glimpse into this issue. In his introductory essay in the book *La connaissance de la vie,* titled "La pensée et le vivant," he concludes with an apt aphorism: "We suspect that, to do mathematics, it would suffice that we be angels. But to do biology, even with the aid of intelligence, we sometimes need to feel like beasts ourselves."[12] The occidental "person-centred" thinking, or—as Paul Rabinow calls it—"subject-oriented approaches"[13] primarily emphasized the highest levels of development in living beings: ego, mature subject, and personhood. The behaviour of the organism was forgotten throughout much of the tradition. From their suggestions it

---

10 Maurice Merleau-Ponty, *Phénoménologie de la perception* (Paris: Gallimard, 1945), 99; *Phenomenology of Perception*, trans. Colin Smith (London/New York: Routledge, 2002 [1962]), 97; henceforth abbreviated as *PP* with French/English page numbers.

11 Colin Smith, *Contemporary French Philosophy: A Study in Norms and Values* (Westport: Greenwood Press, 1976), 129, fn. 1.

12 Georges Canguilhem, *La connaissance de la vie* (Paris: J. Vrin, 2009 [1965; 1952 in Librairie Hachette]), 16; *Knowledge of Life*, trans. Stefanos Geroulanos and Daniela Ginsburg (New York: Fordham University Press, 2008), xx; henceforth abbreviated as *CV* with French/English page numbers.

13 Paul Rabinow, "Introduction: A Vital Rationalist," in Georges Canguilhem, *A Vital Rationalist: Selected Writings from Georges Canguilhem*, trans. Arthur Goldhammer (New York: Zone Books, 1994), 21.

might seem to us that Merleau-Ponty and Canguilhem invert tradition and only speak of the functioning of the organism or animal life in humans. To make this claim however, would be misleading, because both of them emphasize the role of meaning (*sens*) of this functioning. It is not enough to explicate functioning of any organ or of any behaviour, but it is necessary to show the meaning of it.

Merleau-Ponty as well as Canguilhem wrote also about humanity, subjective attitudes, ethical and political claims. While in Merleau-Ponty's case it is clear that he deals more with the notion of subjectivity, in Canguilhem's work it is quite difficult (but not impossible) to find his treatment of subjectivity. The difference between the two approaches could lie in their starting points: The former presupposes the validity of a transcendental philosophy of the ego and of its central aspect in intentionality;[14] the thinking of the latter is at least suspicious of this method of doing philosophy (Canguilhem uses rather the reduced notion of intentionality, and it has never been a starting point for him).[15] For Merleau-Ponty, as presented in *Le visible et l'invisible*, subjectivity has the intentional and bodily characteristics of an attitude to the world ("these visions, these touches, these little subjectivities, these 'consciousnesses of…'").[16] In Canguilhem's case subjectivity is more about what is often termed 'subjective feelings' (to feel somehow subjectively, *se sentir*).[17]

On the following pages I will show even more examples of these intertwined attempts by which their authors philosophically treated certain problems of the

14 Merleau-Ponty, of course, supposes that the ego is "a center from which his intentions radiate." Maurice Merleau-Ponty, *La structure du comportement* (Paris: PUF, 1967 [1942]), 204; *The Structure of Behavior*, trans. Alden L. Fisher (Boston: Beacon Press, 1967), 189; henceforth abbreviated as *SC* with French/English page numbers. And therefore "the objective body is only a moment in the constitution of the object" (*PP,* 86/83).

15 When Canguilhem discusses the problem of 'the corporeal schema' or 'body-schema' (*schéma postural*) he refers not only to Paul Schilder but also to Merleau-Ponty, and he claims that phenomenology of one's own body (*corps propre*) cannot "overcome the paradox of a consciousness of Self as a body in the space." Georges Canguilhem, "Puissance et limites de la rationalité en médecine," in *Etudes*, 408.

16 Maurice Merleau-Ponty, *Le visible et l'invisible* (Paris: Gallimard, 1964), 184; *The Visible and the Invisible*, trans. Alphonso Lingis (Evanston: Northwestern University Press, 1968), 141.

17 Nevertheless, for him these subjective emotions always depend upon an objectively broader horizon of society, nation, history and culture. Cristina Chimisso precisely expresses this relationship as follows: "Fear of illness, or even endurance of pain, can be quite different in different cultures, in which different values are attached to the presence or absence of illness and pain." Cristina Chimisso, "The tribunal of philosophy and its norms: history and philosophy in Georges Canguilhem's historical epistemology," *Studies in History and Philosophy of Biological and Biomedical Sciences* 34 (2003): 320.

living. To begin with, I want to show what it means when Merleau-Ponty and Canguilhem concern themselves with normal behaviour and the normal subject. Next, I will consider the concepts of "the privileged" and "the catastrophic" reactions which both philosophers borrowed from Goldstein. This should lead us to the main theme of this article: dissatisfied life.

2. Is the Normal Subject identical with the Healthy Subject?

In his *La structure du comportement*, Maurice Merleau-Ponty defines the norm through statistical frequency: "By 'norm' here one does not mean a 'should be' which *would make* it be; it is the simple observation of a preferred attitude, statistically more frequent, which gives a new kind of unity to behaviour" (*SC*, 173/159). Hence it seems that for Merleau-Ponty the normal behaviour of anyone follows a frequency of actions signified as the privileged. Is normality of an organism dependent upon the most frequently preferred attitudes in its perception in relation to previous, today less frequent attitudes? Are less frequent attitudes consequently less normal ones? Or must we understand statistical frequency more broadly as a frequency of a type of behaviour in many similar organisms? In fact, these questions belie a misunderstanding. In his statement above, Merleau-Ponty is referring to the conditions in which perception functions. In *Phénoménologie de la perception,* he uses as an example the observation of a table. The colour of the table depends upon the lighting and distance; it is perceived in regards to "'normal' conditions, which mean those which occur most frequently" (*PP*, 351/354). In this case, a subject behaves normally, if the conditions surrounding it are normal, which for Merleau-Ponty means those that appear most often.

However, there is another notion of normality in Merleau-Ponty's work that treats the issue of the abnormal versus normal subject. If we are more sensitive to his other formulations, we are able to notice that Merleau-Ponty uses the standard of normality in contrast with a deficient, ill or blind person's way of perceiving the world. We can demonstrate this through several examples. First, let me consider the following passage from *Phénoménologie de la perception*:

> When we observe that a normal subject is capable of making abstract movements with his eyes shut, and that the tactile experience of the normal person is sufficient to govern motility, it can always be retorted that the tactile data of the normal person have received their objective structure from visual data according to the old conception of the education of the senses. When we observe that a blind person is able to localize *stimuli* on the surface of his body and perform abstract movements—apart from the fact that there are examples of preparatory movements among the blind, the reply can always be made that frequent associations have imparted the qualitative

> colouring of kinaesthetic impressions to tactile ones and welded the former into a quasi-simultaneous occurrence (*PP*, 135–136/134).

The normal person is presented as the opposite of the blind person, and as such the normal person is made identical with the healthy person. We can find similar pairings of opposites in other works. For example, in *La structure du comportement,* we can read that the conditions of patients with visual malfunctions are "removed from the organized perception of the normal person" (*SC*, 72/66). One should not be distracted by the specifics of Merleau-Ponty's examples. What is most important for us to consider lies in the sharp line of demarcation drawn between patient and healthy man, thus between abnormal person[18] and normal person. Before we move on, let me examine another instance of this distinction that Merleau-Ponty makes while giving a description of the normal person: "The so frequent distinction of the mental and the somatic has its place in pathology but cannot serve for the knowledge of normal man, that is, of integrated man, since in him the somatic processes do not unfold in isolation but are integrated into a cycle of more extensive action" (*SC* 195/180).[19] Let me notice that the philosopher speaks exclusively about human beings in this passage, not about animal life in general. He claims that *the normal person is a psycho-somatic and integrated unit.*[20] In contrast to this, *the abnormal being is someone with disintegrated and isolated actions*, not only somebody with malfunctioning 'synaesthetics,' but with a malfunctioning psycho-somatic life, which is not able incorporate all three orders of being—physical, vital and also human order, *and they are not able to coordinate and harmonize them.*

For a clear instance of this kind of lack of perceptual integration we can recall that some schizophrenic people are incapable of connecting their voices with what they intend to say. Merleau-Ponty used a similar example in *La prose du monde*:

> Certain sick people believe that someone else is talking inside their head or in their body, or that someone else is talking when it is they themselves who are pronounc-

18 It seems Merleau-Ponty does not make a distinction between an abnormal person and a person with an anomaly.

19 Compare cited passage with this proposition: "Normal signifies integrated: different character traits are assumed by the total organism." Maurice Merleau-Ponty, "The Child's Relations to Others (1950–1951)," in *Child Psychology and Pedagogy: The Sorbonne Lectures 1949–1952,* trans. Talia Welsh (Evanston: Northwestern University Press, 2010), 279; "Les relations avec autrui chez l'enfant," in *Psychologie et pédagogie de l'enfant. Cours de Sorbonne 1949–1952* (Lagrasse: Verdier, 2001), 350.

20 According to Guillaume Le Blanc, the notion of *intégration* is essential for Merleau-Ponty's understanding of the human living being. Cf. Guillaume Le Blanc, *Canguilhem et la vie humaine* (Paris: PUF, 2010), 80.

> ing or at least mouthing the words. Whatever one's view of the relation between healthy and pathological behaviour, speech must, in its normal functioning, be of such a nature that disorders in it are always possible. There must be something in the very heart of speech which makes it susceptible to these pathologies [*aliénations*].[21]

Thus from all of these examples we are led to understand that the only way to conceive of the perceiving subject is through either the chaos of abnormality or the easily oppressive standard of normal perception. However, as we turn to Canguilhem's treatment of these issues things appear quite different.

At first, Merleau-Ponty's definition of the norm as that which is statistically most frequent is in stark contrast with Canguilhem's thought which critically questioned this understanding of what constitutes a norm.[22] For him, statistical frequency cannot describe the true (qualitative) meaning of norms, as is shown in Canguilhem's dissertation on the normal and the pathological.[23]Also, 'being normal' cannot simply be reduced to 'being healthy'. In other words, normality is not equated with health. Furthermore, in Canguilhem's case the distinction of the normal and the pathological never overlaps with the distinction of the sick[24] and the healthy. That is an essential difference between the approaches of the two philosophers. Now as we move forward let me tease out the implications of this difference.

---

21 Maurice Merleau-Ponty, *The Prose of the World*, trans. John O'Neill (Evanston: Northwestern University Press, 1973), 17; *La prose du monde* (Paris: Gallimard, 2008 [1969]), 26–27. Along these lines we would like to refer to a passage from *La structure du comportement*: "In the same manner, the sentence of a speaker must be organized all by itself, as it were, as happens in fact in the normal use of language in which an awareness of the means of expression for themselves, the contemplation of 'verbal images,' is already a pathological phenomenon" (*SC*, 96–97/87).

22 I would only like to recall that Canguilhem's analyses of normality form a conceptual background to the problem of normality in Husserl's phenomenology in Anthony Steinbock's central passages of his *Home and Beyond*. Cf. Anthony Steinbock, *Home and Beyond: Generative Phenomenology after Husserl* (Evanston, Il: Northwestern University Press, 1995), chs. 8–10.

23 "A human trait would not be normal because frequent, but frequent because normal…" (*NP* 102/160). And later Canguilhem claims that "the concepts of norm and average must be considered as two different concepts" (*NP*, 116/177).

24 In the paper I do not strictly separate the terms of *sickness* and *disease*. However, I would like to refer the reader to the precise differentiations presented by Élodie Giroux between "the medical and objective meaning (*disease*, 'une maladie'), the subjective, existential and clinical meaning (*illness*, 'être malade'), and the subjective and social meaning (*sickness*, 'être un malade')." Élodie Giroux, *Après Canguilhem: définir la santé et la maladie* (Paris: PUF, 2010), 9.

The first things that we have to consider are Canguilhem's reasons for not identifying the normal with the healthy.[25] A simple answer would be as follows: the logic of negation. The strict logical negation of the normal is abnormal, that which does not have or follow a norm. However, there is nothing in the natural world which is without norm. Every living being follows some norm—either higher (superior), or lower (inferior). It is important to note that these lower norms are only "pushed aside by life" (*NP*, 91/144).

If this is true what then is an abnormal living being? In his 1951 essay with the same title as his dissertation "Le normal et le pathologique," Canguilhem explains that if we do not understand the living being and its milieu as being in a reciprocal relationship, then we will be tempted to attribute the label of abnormality, i.e. the pathological to "every anomal individual, every carrier of anomalies—every individual aberrant in relation to a specific, statistically defined type" (*CV*, 208/127). But the aberration is not a negation. Canguilhem often notices that it is the aberrance or distortion of existing norms that is the feature of abnormality rather than the disappearance of norms altogether. "We cannot say that the concept of the 'pathological' is the logical contradictory of the concept 'normal,' for life in the pathological state is not the absence of norms but the presence of other norms" (*CV*, 214/131). More accurately, the French word *anormal* is not the logical contradiction of the French *normal*. In other words, it is not necessary that the *anormal* be understood to be the equivalent of the ideas of 'not-normal', 'being without norm' or 'norm-lessness.' The French prefix *a-* corresponds here rather with the Latin prefix *ab-* which means 'away,' 'from,' 'from out,' rather than 'not.'[26] But if the pathological (the diseased) can be an inferior form of normality, it is not adequate to refer to a normal subject as a healthy subject in opposition to a sick subject, as if the diseased person were devoid of norms or somehow distinguished from the normal one. The sick subject can be normal as well as the healthy one; but the difference lies in the hierarchy and ranking of norms.[27] The healthy subject follows norms with superior

25 "The physiological state is the healthy state, much more than the normal state. It is the state which allows transition to new norms" (*NP*, 155/228). And also: "Inversely, being healthy and being normal are not altogether equivalent since the pathological is one kind of normal" (*NP*, 130/196).

26 See the following etymological passages in *NP*, 81–82/131–133 with these in *CV*, 214/131.

27 For Canguilhem the 'abnormality' of sickness lies in "the incapacity to be normative" (*NP*, 122/186). This incapacity means that the sick person cannot follow their own norms, but only lower, unwanted norms. In short, the normativity is to follow and illimitably create one's own norms.

values; the sick subject follows norms with inferior values—both of them are normal, only governed by different norms.

With all of this in mind now would be a good time to examine a particular statement from Merleau-Ponty:

> Goldstein distinguishes two dimensions of human conduct: the organism attempts simultaneously to maintain life and the essential or functional value of life. In normal subjects the two projects are intermingled; the healthy organism does not pursue mere conservation. In the case of the sick individual, there is a division between the organism's tasks, a division that seeks to ensure the minimum conditions necessary to the pursuit of life; the instinct of conservation appears.[28]

In this passage we are given a clear picture of Merleau-Ponty's effort to valorise rather the normal, healthy and integrated person than the sick one.[29]

3. The 'Narrowed Milieu' and Privileged Behaviour

The production of values is an exclusively human activity and the capacity to follow the highest norms seems often to be a defining characteristic of a healthy person. However, diseases or pathologies place us in other situations in which we are obliged to react differently and to follow inferior norms. Of course, in these situations we do not live a *fully-fledged* life. During a period of disease, a person sometimes reaches their goals through extremely hard effort and frequently by detour, and the person rather chooses for themselves—as Kurt Goldstein calls it—a 'narrowed milieu.'[30] By the term Goldstein means a specific behaviour of living beings when, in a state of anxiety, they create for themselves reduced environments (and consequently expose themselves to reduced self-realizations). The 'narrowed milieu' is a refuge of existence or in Goldstein's own words: "a new plane of 'existence'... *restricted, abnormally concrete world...* This new level is a means to guarantee some kind of existence in a state

28 Merleau-Ponty, "Human Sciences and Phenomenology (1950–1952)," in *Child Psychology and Pedagogy*, 370; "Les sciences de l'homme et la phénoménologie," in *Psychologie et pédagogie de l'enfant*, 462.

29 Let me note that in the fourth from the collection of Merleau-Ponty's radio lectures in 1948, titled "Exploring the World of Perception: Animal Life," he draws another comparison between adult, thus normal and civilized thought and other forms of thought (childish, morbid or barbaric ones). The adult subject is understood as normal, while the childish subject is *at least* insufficiently normal and is at best in the process of becoming normal. Cf. Maurice Merleau-Ponty, *The World of Perception*, trans. Oliver Davis (London and New York: Routledge, 2004), 73.

30 Kurt Goldstein, *The Organism: A Holistic Approach to Biology Derived from Pathological Data in Man* (New York: Zone Books, 1995), 199.

of abnormal relation between the person and the outer world due to the change of the individual by pathology."[31]

The notion of 'narrowed milieu' is considered at length by Canguilhem (*CV*, 215/132) as well as by Merleau-Ponty, who relates it to the effort to conserve one's life, for example, placing oneself in an asylum, into some 'narrow world' (*SC*, 190, fn. 1/245, fn. 97).[32] A concrete example of this restriction of milieu involves the survival instinct. The appearance of the survival instinct is one of the typical characteristics of the sick. The behaviour associated with this instinct involves the subject concentrating all of their activities upon the conservation of life.[33] For Canguilhem a sick person behaves *healthily* if s/he tries to survive in such a 'narrowed milieu,' because s/he follows the 'narrowed' norms of life. S/he would be following a pathological behaviour if s/he always tried to follow only the superior norms of life to which s/he has no access because of his/her sickness. This is, in fact, an interesting paradox of Canguilhem's view: the sick person can behave either 'healthily,' or 'pathologically.'[34] In the former case, the person strenuously tries to adapt to some new, restricted milieu in order to harmonize the disturbed relationship between him/her and the milieu; in the latter, the person anxiously tries to preserve their old and unrestricted milieu, even though it cannot bring them a desired satisfaction.

Canguilhem believes that the problem of the normal and the pathological in Goldstein is focused on the question of the meaning of the phrase *privileged be-*

---

31 Kurt Goldstein, "On Emotions," in *Selected Papers/Ausgewählte Schriften*, ed. Aron Gurwitsch et al. (The Hague: Martinus Nijhoff, 1971), 433.

32 Geroulanos and Ginsburg's English translation uses the expression 'shrunken milieu' for the French *milieu rétréci*; Colin Smith translates it as 'restricted milieu.'

33 That is Goldstein's attitude, too, which we can see in the 1974 article "Organismic Approach to the Problem of Motivation": "Preservation is an expression of pathology, of decay. The goal of the normal organism is not preservation of an existent state. The tendency of normal life is directed toward activity and progress." Nevertheless, several lines above, Goldstein asserts: "The real motive of the activities of the pathological organism is, like that of the normal, the trend to realize its capacities. It tries to realize its preserved capacities in the best possible way." Kurt Goldstein, "Organismic Approach to the Problem of Motivation," in *Selected Papers/Ausgewählte Schriften*, 413–414.

34 "There are two kinds of original modes [*allures inédites*] of life. There are those which are stabilized in new constants but whose stability will not keep them from being eventually transcended again. These are normal constants with propulsive value. They are truly normal by virtue of their normativity. And there are those which will be stabilized in the form of constants which the living being's every anxious effort will tend to preserve from every eventual disturbance. These are still normal constants but with repulsive value expressing the death of normativity in them. In this they are pathological, although they are normal as long as the living being is alive" (*NP*, 137/206).

haviour (*comportement privilégié*).[35] Privileged behaviour is again a notion which we can find in the writings of both these French authors. Let me start by comparing two of the following quotations. To begin, we can read in Merleau-Ponty: "For the most part preferred behaviour is the simplest and most economical *with respect to the task in which the organism finds itself engaged;* and its fundamental forms of activity and the character of its possible action are presupposed in the definition of the structures which will be the simplest *for it,* preferred *in* it" (*SC*, 159/147). Now we can pass to the reading of Canguilhem's suggestion: "One cannot understand its [organism's] actions without appealing to the notion of privileged behaviour. 'Privileged' does not mean objectively simpler—just the inverse. The animal finds it simpler to do what it privileges. It has its own vital norms" (*CV*, 187/113).[36]

So, what does privileged behaviour mean? What is the relation between the simple and the privileged? Both Merleau-Ponty and Canguilhem derive their conceptions of privileged behaviour from the work of Goldstein, for whom such behaviour is "subjectively characterized by the feeling of comfort, agreeableness, security, and correctness."[37] It seems that Merleau-Ponty's views are closer to such an explanation, because for him privileged behaviour is built around the standard of what is optimal for the organism. That said, it would not be correct to say that Canguilhem misunderstood Goldstein's proposition. We would like to emphasize that Canguilhem speaks of the privileged as the 'objectively' not-simple.

The problem with privileged behaviour is linked to the problem of stimuli in experimental conditions. Privileged behaviour is the typical reaction of an organism to the given stimuli in an artificial milieu. However, it could seem to us that Merleau-Ponty identifies it with simplicity and optimality:

> For the preferred behaviour is the one which permits the easiest and most adapted action: for example, the most exact spatial designations, the finest sensory discriminations. Thus each organism, in the presence of a given milieu, has its optimal conditions of activity and its proper manner of realizing equilibrium; and the internal

---

35 See Canguilhem's manuscript titled "Normalité et normativité," in *A Vital Rationalist*, 354–355. In most cases we will use for the French *privilégié* the English word *privileged*, even if occasionally we run into the words *preferred* or *preferential* in certain translations.

36 A different version is found in John Savage's English translation: "We cannot understand its action without thinking of it in terms of a privileged form of behaviour. 'Privileged' does not mean objectively simpler in this case; it is rather the opposite. The animal finds it easier to do what it favors: it follows its own norms of living." Georges Canguilhem, "Living Being and its Milieu," trans. John Savage, *Grey Room* 2 (2001): 21.

37 Cf. Goldstein, *The Organism*, 273.

> determinants of this equilibrium are not given by a plurality of vectors, but by a general attitude toward the world (*SC*, 160–161/148).

Nevertheless, it is not so easy to identify the privileged with what seems objectively most simple, if we read another passage in the text (very similar to Canguilhem, as we can see above): "It is not because behaviour is simple that it is preferred; on the contrary it is because it is preferred that we find it simpler" (*SC*, 159/147).[38]

It seems that Merleau-Ponty understands optimality as a sign of 'privileged position' in perception, as we can find in *Phénoménologie de la perception* (cf. *PP* 349/352). Merleau-Ponty concentrates his effort on studying how an organism perceives or experiences a situation from its subjective point of view. For him it is a matter of the optimal setting of a subjective attitude to the milieu or to the world.

It is quite different in Canguilhem's case. His project causes him to understand privileged behaviour as the normative conduct of healthy individuals. In *Le normal et le pathologique* Canguilhem defines privileged behaviour as follows:

> By preferential behaviour it must be understood that among all the reactions of which an organism is capable under experimental conditions, only certain ones are used as those preferred. This mode of life, which is characterized by a set of preferred reactions, is that in which the living being responds best to the demands of his environment and lives in harmony with it; it is that which includes the most order and stability, the least hesitation, disorder and catastrophic reactions (*NP*, 121/184–185).

Thus, not optimality in position, but rather *organismic quietness* (as a kind of 'unconscious' behaviour) would be a more adequate attribute for privileged behaviour. Canguilhem frequently quoted René Leriche's sentence, that health is "life lived in the silence of the organs" (e.g. *NP*, 52, 59, 72, 180/91–92, 101, 118, 243; *CV*, 211/129). A healthy people feel they are in harmony with their milieu, and therefore are not aware of their health. A disorder or disharmony must take place in the person's consciousness in order to make them aware of

38 Compare it with Goldsteins's claim which was interesting to Merleau-Ponty: "But what is simple? That certainly cannot be determined on the basis of the content of the process or of the experience. The determination will depend on whether one departs from isolable 'part contents' or from the whole. By the fact that something is 'a figure,' a Gestalt, it apparently becomes simple for us. This probably also holds true with regard to physiological processes. But this is precisely the problem: Why is something a Gestalt? Simplicity and pregnanz, after all, are nothing but essential characteristics of a Gestalt. But what causes a Gestalt? This is what we really wish to know." Goldstein, *The Organism*, 275.

their unsoundness and weaken their ability to act in a common, privileged manner. In short, to act preferentially is to live one's own life in unconscious welfare.

If there is no awareness or anxiety resulting from dangerous scenarios concerning life or survival, the organism does not become aware of the need to react quickly against dangerous stimuli. In his lecture "Le vivant et son milieu," Canguilhem explains it by using a very instructive example from the work of von Uexküll concerning a tick. An adult female tick is able to lay in wait for up to eighteen years, if there is no mammal in its milieu. The stimulator for it is the smell of rancid butter. If we prepare it artificially, the tick drops down, but it does not attach to the 'illusory body' because it contains no blood. The tick climbs up its tree and it again waits for prey. When a real mammal stays under the tree, the tick drops down and sucks its blood. "Only at the moment when the mammal's blood enters into her stomach do the tick eggs (encapsulated ever since the moment of mating and capable of remaining encapsulated for eighteen years) open, mature, and develop. The tick can live for eighteen years in order to perform her reproductive function in just a few hours" (*CV*, 186/112). The waiting and the dropping down are the simplest ways (minimum of energy, economical conduct) of achieving the privileged behaviour, which for it is nutrition and reproduction. In this case simple means neither fast nor optimal; quickness and optimality are human, anthropomorphic expectations.

Canguilhem shows us that it is not the tick's passive and long-term waiting or effortless dropping down that characterize its privileged behaviour. Waiting and dropping are indeed simpler, but privileged behaviour is the whole of these reactions, a style of life in harmony (or more precisely, in equilibrium) with the milieu. It is necessary to notice that the privileged behaviour presupposes one condition: a healthy subject. Why? Because the healthy subject is able to perform all possible acts and one of them is privileged. A sick subject does not have access to all of the possibilities. The subject's possibilities are limited. With a sick subject, this privileged behaviour is achieved only with great difficulty, more difficulty at least than in the case of a healthy subject. As we have seen above, health and normalcy are not the same things and that is the reason why Canguilhem can assert: "Man is truly healthy only when he is capable of several norms, when he is more than normal" (*CV*, 215/132). Achieving equilibrium with one's milieu is a perpetual process. As Goldstein says, the organism cannot once and for all find the adequate milieu.[39] In this same vein, for Canguilhem,

39 "The organism never lives in a completely adequate milieu but must continuously assert itself against inadequate stimuli, that is, against such stimuli that evoke isolated effects." Goldstein, *The Organism*, 227.

health "is not a constant of satisfaction, but the *a priori* power to master perilous situations, this power uses itself up in mastering successive perils."[40]

4. Conclusions

In my introduction I wrote that in order to do phenomenology it is not important whether or not Georges Canguilhem is a phenomenologist. It is also not important that he never analyzed Husserl's major works and works of Husserlian thinkers. If he belongs—as Foucault suggests—to the camp of philosophers such as Cavaillès, Koyré and Bachelard, this is based not only on their mode of philosophical work, but also on the modality of doing phenomenology which is, according to him, a different one from those of Merleau-Ponty or Sartre. Of course, I admit that there is heterogeneity of the two camps in terms of their thoughts; they have different roots in philosophical training and in ways of doing philosophy, but there is a common inspiration which arises—explicitly, or implicitly; directly, or indirectly—from Husserlian phenomenology.

Foucault differentiates the mentioned camps of philosophers from the only perspective, and that is why Merleau-Ponty and Sartre belong to the same group. Let me suggest at least in brief another possible perspective. If we read in Bachelard's *La terre et les rêveries de la volonté* about conflict between phenomenology of '*vers*' and phenomenology (or more precisely, dynamology) of '*contre*,'[41] we can see it from a different point of view. While phenomenology of '*vers*' is in reality Sartre's visual phenomenology of object, phenomenology of '*contre*' is attributed directly to Merleau-Ponty and it is closer to Bachelard's own phenomenology of matter.[42] By "classical phenomenology"[43] Bachelard means mostly Sartre's passive phenomenology of intentionality directly oriented to (*vers*) an object. He puts this phenomenology in contrast to 'controlled phenomenology' (*la phénoménologie dirigée*), which is more dynamic and in which consciousness is named as tenacious or working with effort (*la conscience opiniâtre*) against (*contre*) the matter.[44] Bachelard's distinction of two modalities of phenomenology is based evidently on a different point of view than that of Fou-

40 Georges Canguilhem, "Is a Pedagogy of Healing Possible?" trans. Steven Miller, *Umbr(a): Incurable* 10 (2006): 19.

41 Gaston Bachelard, *La terre et les rêveries de la volonté: essai sur l'imagination de la matière* (Paris: José Corti, 2004 [1948]), 57–58.

42 Gaston Bachelard, *Le matérialisme rationnel* (Paris: PUF, 1963 [1953]), 37; and also Bachelard, *La terre et les rêveries de la volonté*, 112–116.

43 Bachelard, *Le matérialisme rationnel*, 11.

44 I broadly discussed such distinction in my book *Gaston Bachelard a filozofia pohľadu* (Trnava: Filozofická fakulta TU, 2012), 82–84.

cault, and from this perspective Merleau-Ponty and Sartre will belong to different groups. Bachelard makes some necessary rectifications in the common understanding of phenomenology and its role in the sciences in order to ensure that phenomenology can be "a contemporary of science"[45] and to be scientific itself. And it is obvious that in certain respects he sees Merleau-Ponty's phenomenology as being closer to contemporary science than Sartre's.

In terms of their phenomenological aspects, the writings of these two philosophers are interesting in various ways. For both of them, biological issues (problems of organism, of living being, of animality) are, as it were, a *terra incognita* of philosophical thinking. Merleau-Ponty often suggested that a Cartesian heritage prevents philosophy from developing this issue much beyond the frame of subject and object correlation. In his lectures he emphasized that an organism is not a *constructum*;[46] and in his commentary to Husserl's *The Crisis* he also asserted that an organism is not a *bloße Sachen* (a mere thing), but *Einfühlung* (empathy) and as such "it is not a residue of pre-science in science—as [the] Cartesian ontology believes—; it is [the] proximity of 'sources' (i.e. of total image of the world in *Ineinander*), proximity of *Tiefe* which means that biology always leads to transcendental questions, always makes an incursion into the terrain of *Lebenswelt*."[47]

Of course, transcendental questions play no role in Canguilhem's approaches to biological issues; he was not a philosopher of transcendentalism or transcendental phenomenology. It is not so easy to use Foucault's distinction of two modalities of the reception of phenomenology in France, and in consequence to claim that there are also two strictly different kinds of phenomenology in France: a transcendental phenomenology (Sartre, Merleau-Ponty) on the one hand, and a non-transcendental (i.e. mere mundane) phenomenology (Bachelard, Canguilhem) on the other hand. That is to say, Merleau--Ponty overcomes that distinction and he develops both kinds (when the latter is a leading clue to the former). And if we want, nonetheless, to see some assertions very close to phenomenology in Canguilhem's writings, it seems we might describe him as a non-transcendental phenomenologist, though I suppose he would never agree with such labelling.

Above all, we could state that Georges Canguilhem is a thinker of dissatisfaction. But what does this mean? To answer this, I would refer to several pas-

45 Georges Canguilhem, "L'histoire des sciences dans l'œuvre épistémologique de Gaston Bachelard," in *Etudes*, 191.

46 Maurice Merleau-Ponty, *La nature. Notes. Cours du Collège de France* (Paris: Éditions du Seuil, 1995 [1968]), 117.

47 Merleau-Ponty, *Notes de cours 1959–1961*, 90.

sages from his writings. For Canguilhem, the theme of dissatisfaction became a point of concern long before his dissertation on the normal and the pathological. In his younger years, he wrote two texts in which he spoke of two different kinds of philosophy. It was with these two types of philosophy in mind that he gave a speech in 1926 titled "Ce que pense la jeunesse universitaire française," in which he claims that it is necessary to distinguish between intellectuals who are just coming and those who have already arrived.[48] These two groups of intellectuals fit clearly into a very significant paradigm in his thought. This paradigm shows itself again in another text in which he suggests that there are two different facets to the history of philosophy: there is the philosophy of 'already' [*déjà*] and the philosophy of 'still' [*encore*].[49] We can talk about philosophy which has achieved everything that it wanted, and about philosophy that is never satisfied by what can be achieved. It is this insight that could be the first expression of the idea of dissatisfaction that forms the basis of Canguilhem's philosophical heritage.

We have seen that for Canguilhem it is not necessary to conceive of sickness as merely passive—as a state of conservation of life. It is true that the sick person has some abilities, even if they are limited: the sick person is able to fight against disease; in such a situation, resignation would be a manifestation of death. Only in death does a person cease striving. Canguilhem inscribed into his works a restless struggle against death. A clear example of this is in the inspiring article "Is a Pedagogy of Healing Possible?" Here he notices that the healthy person must be open to the possibility of becoming sick.[50] Let me quote his—literally—Heideggerian expression: "Human life is an existence, a being-there for a non-preordained becoming haunted by its own end. The human being is thus exposed to disease, not by some condemnation or destiny, but by his simple presence in the world."[51] After these words Canguilhem discusses the topic of Goldstein's *Der Aufbau des Organismus* which was accepted only by "philosophical circles influenced by the work of Maurice Merleau-Ponty."[52] It might be surprising that Canguilhem—even if indirectly— would refer to Heidegger, but it seems less surprising if one sees that the influence comes at least partially

48 "A cela il n'y a qu'une remedè: être toujours *en train de venir* et mépriser *ceux qui sont arrives*." Georges Canguilhem, "Ce que pense la jeunesse universitaire française," in *Œuvres complètes 1: Écrits philosophiques et politiques 1926–1939* (Paris: J. Vrin, 2011), 151 [my emphasis].

49 Cf. Georges Canguilhem, "Une conception récente du sentiment religieux et du sacré," in *Œuvres complètes 1*, 320.

50 In *Le normal et le pathologique*, he calls it 'a biological luxury' (*NP*, 132/199).

51 Canguilhem, "Is a Pedagogy of Healing Possible?" 15–16.

52 *Ibid.*, 16.

through Goldstein, who directly recalls Heidegger in his work when he compares catastrophic reactions of organism with anxiety.[53] A consideration of anxiety appears also in Canguilhem's writings, even if it is not precisely in Heidegger's vein of thinking. Canguilhem's unique understanding of the problem of anxiety is expressed in at least four ways: (1.) by basic disquiet, (2.) by the polarity of life, (3.) by the notion of transgression, and finally (4.) by fragility.[54]

First, the meaning of disquiet can be seen in the following sentence: "Knowledge, then, is an anxious quest [*une recherche inquiète*] for the greatest possible quantity and variety of information. If the *a priori* is in things, if the concept is in life, then to be a subject of knowledge is simply to be dissatisfied with the meaning one finds ready at hand [*du sens trouvé*]. Subjectivity is therefore nothing other than dissatisfaction. Perhaps that is what life is."[55] A totally satisfied life would be equal to death. For Canguilhem, fixed health never exists; it is a contradiction. To be absolutely healthy is to cease being a living person. Living, or more generally, life is essentially in conflict with the static and the closed.[56]

If thinking and life are not in opposition, then philosophy can be understood as an expression of life, and we can agree with Thomas Osborne who characterizes Canguilhem's notion of philosophy as "tentative, open, ever-revisable, inherently problematological."[57] For him thinking is like life: never settled and never a closed subject.[58]

---

53 Cf. Goldstein, *The Organism*, 231. Guillaume Le Blanc also refers to Goldstein's particular note about Heidegger. Cf. Le Blanc, *Canguilhem et la vie humaine*, 77, fn. 2. Let me note that in his article "On Emotions," Goldstein connects the problem of anxiety with Jean-Paul Sartre's thinking. cf. Goldstein, "On Emotions," 432–434.

54 We can at least partially find here some similarities with Heideggerian thinking about anxiety when *Dasein* stays face to face with death. As Jana Tomašovičová suggests, considering the notion of *reservation*: the loneliness "leads to the reservation that Dasein got tangled up in himself. Such reservations are reinforced by the fact Dasein's authentic being itself culminates in its confrontation with its own finitude." Jana Tomašovičová, "Existential Analytics and the Social Sciences," in *Affectivity, Agency and Intersubjectivity*, ed. Peter Šajda et al. (Budapest: L'Harmattan, 2012), 101.

55 Canguilhem, "La nouvelle connaissance de la vie," 364; *A Vital Rationalist*, 319.

56 For Canguilhem, life is always an "activity of opposition to inertia and indifference," thus against death (*NP*, 173/236).

57 Thomas Osborne, "What is a Problem?" *History of the Human Sciences* 16 (2003): 10.

58 Nikolas Rose says about Canguilhem's history of reason that it does not have "a linear and progressive form, it is not a progress to truth." Nikolas Rose, "Life, reason and history: reading Georges Canguilhem today," *Economy and Society* 27 (1998): 159. Let me notice that even though affinity does not mean identity, and also while the absence of

Therefore, the reason of the vitality—and this is the second form of anxiety mentioned above—lies in his understanding of life as characterized by polarity, or more precisely, by *dynamical polarity*.[59] Polarity expresses the *polemos*, the struggling between negative and positive values, "because by taking values into consideration one arrives at the most originary core of the personality" (*CV*, 162/96). The polarized dissatisfaction of life and of thinking also has an element of heroism, one which Élisabeth Roudinesco attributes to the whole of Canguilhem's political life and his philosophy of science.[60] We can find in Bachelard the roots of this topic, as well. Canguilhem refers to him in "Nouvelle réflexions concernant le normal et le pathologique," when he recalls that Bachelard was interested in the dynamical opposition of values and anti-values (*NP*, 176–177/239).

A concrete expression of this concept of vital struggling exists in Canguilhem's political philosophy in the form of transgression (*infraction*). Social and political rules are like a milieu; they should not be eternally fixed, but open to revisions. And for Canguilhem this is the reason why transgression should be conceived as the very beginning of the regulation process (*NP*, 179/242). As in the case of health, norms can be either superior or inferior, either progressive or regressive, either open or closed. In the manuscript "Normalité et normativité," Canguilhem writes:

> Any normality open to possible future correction is authentic normativity, or health. Any normality limited to maintaining itself, hostile to any variation in the themes that express it, and incapable of adapting to new situations is a normality devoid of normative intention. When confronted with any apparently normal situation, it is therefore important to ask whether the norms that it embodies are creative norms,

---

a term is not the absence of a notion, the concept of approximation to truth is presented in Canguilhem's writings. This idea is supported by "Traité de logique et de morale" written by Canguilhem and Camille Planet, where we can see it as essential in this statement: "The personality of the scientist enters into the science. The science is a personal effort towards impersonal truth." Georges Canguilhem and Camille Planet, "Traité de logique et de morale," in Canguilhem, *Œuvres complètes 1*, 698.

59 See for example *NP*, 84–85/135–136.

60 Cf. Elisabeth Roudinesco, *Philosophy in Turbulent Times: Canguilhem, Sartre, Foucault, Althusser, Deleuze, Derrida*, trans. William McCuaig (New York: Columbia University Press, 2008), 1–32. As an excellent basic introduction to Canguilhem's political thinking we can recommend Jacques Bouveresse's preface to the first volume of Canguilhem's *Œuvres complètes*, especially the chapter titled "Du pacifisme à la Résistance: le rôle d'Alain dans la formation de Canguilhem." Jacques Bouveresse, "Préface aux *Œuvres complètes* de Georges Canguilhem," in Canguilhem, *Œuvres complètes 1*, 30–40.

> norms with a forward thrust, or, on the contrary, conservative norms, norms whose thrust is toward the past.[61]

There is one more aspect of the subject of anxiety: the precarious situation of the living being, its fragility. It supposes a polemical characterization—the possibility of destruction. For Canguilhem, a typical example of fragile existence is a cell, but also illusorily stable 'substances' are in fact exposed to destructive forces. Knowledge emerged as one of the ways of overcoming and at least partially stabilising this fragility.

Thus, if knowledge arose in human history from the dissatisfaction of a living being with a given milieu, if knowledge cannot be understood only as an absurd 'catechism' of eternally valid and fixed scientific truths, (which was the central idea of Bachelard's as well as Canguilhem's criticism of positivistic approaches to history of science) and if knowledge is a normative activity, then we can attribute the same attitude to Merleau-Ponty, who characterized one's own research as one's own 'interior disorder'[62] and in a noteworthy reference to Husserl he remarked:

> Philosophy [is not] beyond time and place; philosophy is situated in the *lebendige Gegenwart*, i.e. Being is vertical, it is always in my *Präsenz-Feld* in the manner of '*dunkles Wissen*' and not of definition or *Wortbedeutung*—'transcendental Life'—open on principle. It is a mistake to believe that philosophy is the ideas. Rather, it is a field of questioning which does not know what about itself it asks (philosophy is a problem for the philosopher himself).
>
> [...]
>
> Philosophy becomes philosophy of philosophy; it considers itself as a *Gebilde* and it ceases to be a simple *Gebilde*, when it takes a birth in 'philosophical life': in my present living, there is an overcoming of 'wandering' [*baladeuse*] philosophy, which will again descend into historical relativity and the *Lebenswelt*. And what overcomes the historical relativity and is the seat of philosophy, that is the self and the other philosophers conceived as the living minds (and not in something timeless); that is the mind which is neither as an objective mind (history process), nor as a subjective mind (that I know about philosophy, my definitions, my 'consciousness'), but the mind as the *Ineinander* of philosophies one in another—their common root: a pure questioning.[63]

---

61 Canguilhem, "Normalité et normativité," in *A Vital Rationalist*, 352.

62 Cf. Maurice Merleau-Ponty, "Éloge de la philosophie," in *Éloge de la philosophie et autres essais* (Paris: Gallimard, 2011 [1953]), 13.

63 Merleau-Ponty, *Notes de cours 1959–1961*, 86–87.

In reality, all dissatisfied questioning in the living is the very origin of engaging in philosophy; and all philosophy is possible only if a person encounters a conflict somewhere in their world.[64]

References

Bachelard, Gaston. *La terre et les rêveries de la volonté: essai sur l'imagination de la matière.* Paris: José Corti, 2004 [first published 1948].

— *Le matérialisme rationnel.* Paris: PUF, 1963 [first published 1953].

Canguilhem, Georges. "Is a Pedagogy of Healing Possible?" Translated by Steven Miller. *Umbr(a): Incurable* 10 (2006): 9–21.

— "Living Being and its Milieu." Translated by John Savage. *Grey Room* 2 (2001): 6–31.

— *A Vital Rationalist: Selected Writings from Georges Canguilhem.* Translated by Arthur Goldhammer. New York: Zone Books, 1994.

— *Etudes d'histoire et de philosophie des sciences concernant les vivants et la vie.* Paris: J. Vrin, 2002 [first published 1968].

— *Idéologie et rationalité dans l'histoire des sciences de la vie.* Paris: J. Vrin, 2009 [first published 1977].

— *Knowledge of Life.* Translated by Stefanos Geroulanos and Daniela Ginsburg. New York: Fordham University Press, 2008.

— *La connaissance de la vie.* Paris: J. Vrin, 2009 [first published 1962; 1952 in Librairie Hachette].

— *Le normal et le pathologique.* Paris: Quadrige/PUF, 2011 [first published 1966].

— *Œuvres complètes 1: Écrits philosophiques et politiques 1926–1939.* Paris: J. Vrin, 2011.

— *The Normal and the Pathological.* Translated by Carolyn R. Fawcett. New York: Zone Books, 1991 [first published 1989].

— *Vie et mort de Jean Cavaillès.* Paris: Allia, 2004 [first published 1976].

Chimisso, Cristina. "The tribunal of philosophy and its norms: history and philosophy in Georges Canguilhem's historical epistemology." *Studies in History and Philosophy of Biological and Biomedical Sciences* 34 (2003): 297–327.

Derrida, Jacques. "Introduction" to *L'origine de la géométrie* by Edmund Husserl, 3–171. Paris: PUF, 2010 [first published 1962].

Foucault, Michel. "La vie: l'expérience et la science." *Revue de métaphysique et de morale* 90 (1985): 3-14.

— "Life: Experience and Science." Translated by Robert Hurley. In *Essential Works of Foucault 1954–1984 (Aesthetics, Method, and Epistemology)*, Volume II, edited by Paul Rabinow, 465–478. New York: The New Press, 1998.

Giroux, Élodie. *Après Canguilhem: définir la santé et la maladie.* Paris: PUF, 2010.

64 The article was produced at the Department of Philosophy of Trnava University as a part of the VEGA 1/0272/13 grant project "Interpersonality, Normativity, Generativity: Phenomenological Research."

Goldstein, Kurt. "The Organismic Approach to the Problem of Motivation." In *Selected Papers/Ausgewählte Schriften*, edited by Aron Gurwitsch, Else M. Goldstein Haudek, and William E. Haudek, 409–421. The Hague: Martinus Nijhoff, 1971.

— *The Organism: A Holistic Approach to Biology Derived from Pathological Data in Man*. New York: Zone Books, 1995.

Le Blanc, Guillaume. *Canguilhem et la vie humaine*. Paris: PUF, 2010.

Lipták, Michal. "Raný Merleau-Ponty v diskusii s naturalizmom a psychológiou." *Filozofia* 66 (2011): 935–943

Merleau-Ponty, Maurice. *Éloge de la philosophie et autres essais*. Paris: Gallimard, 2011 [first published 1953].

— *Child Psychology and Pedagogy: The Sorbonne Lectures 1949–1952*. Translated by Talia Welsh. Evanston: Northwestern University Press, 2010.

— *La nature. Notes. Cours du Collège de France*. Paris: Éditions du Seuil, 1995 [first published 1968].

— *La prose du monde*. Paris: Gallimard, 2008 [first published 1969].

— *La structure du comportement*. Paris: PUF, 1967 [first published 1942].

— *Le visible et l'invisible*. Paris: Gallimard, 1964.

— *Notes de cours 1959–1961*. Paris: Gallimard, 1996.

— *Phénoménologie de la perception*. Paris: Gallimard, 1945.

— *Phenomenology of Perception*. Translated by Colin Smith. London/New York: Routledge, 2002 [first published 1962].

— *Psychologie et pédagogie de l'enfant. Cours de Sorbonne 1949–1952*. Lagrasse: Verdier, 2001.

— *The Prose of the World*. Translated by John O'Neill. Evanston: Northwestern University Press, 1973.

— *The Structure of Behavior*. Translated by Alden L. Fisher. Boston: Beacon Press, 1967.

— *The Visible and the Invisible*. Translated by Alphonso Lingis. Evanston: Northwestern University Press, 1968.

— *The World of Perception*. Translated by Oliver Davis. London and New York: Routledge, 2004.

Osborne, Thomas. "What is a Problem?" *History of the Human Sciences* 16 (2003): 1–17.

Rose, Nikolas. "Life, reason and history: reading Georges Canguilhem today." *Economy and Society* 27 (1998): 154–170.

Roudinesco, Elisabeth. *Philosophy in Turbulent Times: Canguilhem, Sartre, Foucault, Althusser, Deleuze, Derrida*. Translated by William McCuaig. New York: Columbia University Press, 2008.

Smith, Colin. *Contemporary French Philosophy: A Study in Norms and Values*. Westport: Greenwood Press, 1976.

Steinbock, Anthony. *Home and Beyond: Generative Phenomenology after Husserl*. Evanston, Il: Northwestern University Press, 1995.

Tomašovičová, Jana. "Existential Analytics and the Social Sciences." In *Affectivity, Agency and Intersubjectivity*, edited by Peter Šajda et al., 96–110. Budapest: L'Harmattan, 2012.

Vydra, Anton. *Gaston Bachelard a filozofia pohľadu*. Trnava: Filozofická fakulta TU, 2012.

—

Anton Vydra
Center for Phenomenological Studies
Department of Philosophy
Faculty of Philosophy and Arts
Trnava University in Trnava
Hornopotočná 23
918 43 Trnava
Slovak Republic
tonovydra@gmail.com

Heideggers Humanismuskritik und der gegenwärtige Transhumanismus

Jana Tomašovičová

Abstract

As a key issue, human enhancement nowadays shapes many philosophical and scientific discussions. Controversy largely springs from concerns about the moral acceptability of human gene pool manipulation for the purposes of enhancing humanity by improving its physical, cognitive and emotional capabilities. However, whether mankind will benefit from such actions is a question which still needs to be assessed unequivocally. This article therefore analyzes the reasoning behind the main critique—with special emphasis on Heidegger's critique—directed at present humanistic theories in order to investigate whether the most disputable points that have been a part of humanistic projects, and as such have been legitimately criticized, are not a part of transhumanistic visions as well. The article also addresses the questions of whether Heidegger, besides his critique of humanism, also offers a concrete solution that could show us how to eliminate its problematic aspectstand to what extent his investigations of humanity are able to extend contemporary discussions about transhumanistic tendencies.

Keywords: humanitas, animal rationale, ek-sistence, transhumanism, genetic manipulation

Die Sachen, an denen etwas ist, kommen,
auch wenn sie nicht für die Ewigkeit bestimmt sind,
selbst in spätester Zeit noch rechtzeitig.[1]

Martin Heideggers Humanismuskritik, die auf die Korrespondenz zwischen ihm und dem französischen Denker Jean Beaufret zurückgeht und 1947 unter dem Titel *Brief über den ‚Humanismus'* publiziert wurde, fand eine bemerkenswerte Resonanz in den 60er und 70er Jahren des 20. Jahrhunderts, insbesondere im Umfeld der französischen Philosophie. Sie regte eine große Welle von Auseinandersetzungen an, in denen es um die Grenzen des Humanismus und Antihumanismus ging und die im Geiste der Subjekt- und Anthropozentrismuskritik ausgetragen wurden. Sie nahm an der Gestaltung mehrerer strukturalistischer Projekte und postmoderner Visionen teil und beeinflusste vor allem die Überlegungen über den Menschen bei M. Foucault, J. Derrida oder J.-F. Lyotard. Heutzutage können die kritischen, sich auf den Humanismus beziehenden Reflexionen des späten Heidegger wieder einmal ernst genommen und erneut in den philosophisch-wissenschaftlichen Diskurs einbezogen werden, in dem dies-

1 Martin Heidegger, "Brief über den 'Humanismus,'" in *Wegmarken*, Gesamtausgabe Bd. 9, ed. Friedrich-Wilhelm von Herrmann (Frankfurt am Main: Vittorio Klostermann, 1976), 344.

mal über das zukünftige Erscheinungsbild des Menschen und seine neuen Fähigkeiten, Verbesserungen und Modifizierungen, die mit Hilfe von sich schnell entwickelnden modernen Technologien und Genmanipulationen zustande kommen, entschieden wird. Mit seinen kritischen Betrachtungen über den Menschen ist es Heidegger bereits gelungen, den etablierten traditionellen Philosophiekonzepten den Spiegel vorzuhalten. Deshalb möchten wir in diesem Beitrag zuerst untersuchen, welche Hauptgründe Heidegger zur Humanismuskritik bewogen haben und ob er auch konkrete Lösungsansätze vorschlägt, wie man die problematischen und aus seiner Sicht strittigsten Punkte der humanistischen Konzeptionen eliminieren kann. Aufgrund dieser Analysen stellen wir uns anschließend die Frage, inwieweit Heideggers Untersuchungen zur Humanität und der Wesensbestimmung des Menschen die gegenwärtigen Diskussionen über transhumanistische Tendenzen, bei denen der Ernst und das Verantwortungsbewusstsein des philosophischen Denkens auf radikale Weise immer dringlicher werden, erweitert und vertieft werden können. Dabei überprüfen wir auch Heideggers These aus dem Anfangszitat, ob wir überhaupt in der Lage sind, die auftauchenden wesentlichen Themen und Probleme zu erkennen, und ob zu diesen in der gegenwärtigen historischen Situation auch das Erfassen des menschlichen Wesens gehört.

1. Beweggründe für Heideggers Humanismuskritik

Der Begriff Humanismus fand den Anfang seiner Anwendung erst Ende des 19. Jahrhunderts, die Wurzeln der sprachverwandten Ausdrücke *humanitas* und *studia humanitatis* sind jedoch viel älter und reichen in die Zeit der römischen Republik zurück.[2] Nach Heidegger hatten schon die alten Römer, das griechische Kulturgut aufgreifend, die Bemühung um *humanitas*, also darum, dass der Mensch menschlich wird, vor allem mit Bildung und dem Studium griechischer Literatur, die das menschliche Wesen veredeln, vervollkommnen und erziehen sollten, in Verbindung gebracht. Der spätere Humanismus der Renaissance weckte im Gegensatz zur mittelalterlichen scholastischen Lehre ein erneutes Interesse an der antiken Kultur, vor allem an griechischen und lateinischen literarischen Werken, deren Helden oft vorbildhafte menschliche Charaktere und Einstellungen verkörperten. Mittels dieser spezifischen Art der Bildung, mittels *studia humanitas* und weiterer Kultur- und Bildungsaktivitäten wollte auch der Renaissancehumanismus vor allem zur Erziehung, Entfaltung und Erhabenheit des Menschen beitragen. Heidegger konstatiert deshalb, dass sich die unterschiedli-

2 *GA* 9, 320. Vgl. auch Paul Oskar Kristeller, *Der italienische Humanismus und seine Bedeutung* (Basel: Helbing & Lichtenhahn, 1969).

chen historischen Ausprägungen des Humanismus insgesamt darum bemühen, dass der Mensch für seine Menschlichkeit frei werde und darin seine Würde finde.[3] Eine solche Vision blieb wohl niemandem fremd, weshalb nach Dominique Janicaud „im 20. Jahrhundert fast alle Denkschulen Humanismus forderten, sowohl Materialisten als auch Idealisten".[4] Es war also nur eine Frage der Zeit, wann das Wort Humanismus infolge der vagen Anwendung seine Bedeutung verliert. Genau das hatte auch J. Beaufret in seiner Frage an Heidegger reflektiert,[5] wobei er Heidegger zum Nachdenken über Humanismus und zur Kritik des humanistischen Projektes veranlasste.

Da die Bemühung um *humanitas* ursprünglich eine Bemühung um Menschlichkeit war, also Bemühung um das Verstehen dessen, was den Menschen menschlich macht und was seinen spezifischen Charakter ausmacht, stellte sich Heidegger anfangs die Frage, *woher und wie* überhaupt das Denken über das Wesen des Menschen in den verschiedenen historischen Arten des Humanismus bestimmt wurde. Es ist nämlich ein Unterschied, ob der Mensch nur gegen das andere vorhandene Seiende abgegrenzt und also in der Linie der metaphysischen Seinsvergessenheit interpretiert wird oder ob man das Wesen des Menschen aufgrund seines Bezugs zum Sein denkt. Nach Heidegger kommen die einzelnen Arten des Humanismus darin überein, dass in ihrem Menschenbild die Menschlichkeit „aus dem Hinblick auf eine schon feststehende Auslegung der Natur, der Geschichte, der Welt, des Weltgrundes, das heißt des Seienden im Ganzen bestimmt wird."[6] Obwohl Heidegger hier auf eine gewisse Verallgemeinerung der historischen Arten des Humanismus nicht verzichten konnte, liegt der Schlüsselpunkt seiner Überlegungen in der Einsicht, dass die humanistischen Versuche um das Erfassen des Menschen und um seine Wesensbestimmung jeweils in einem bestimmten vorgegebenen Interpretations- und Auslegungsrahmen geschahen. Diesen dominierenden Rahmen stellte das metaphysische Denken dar. Jede Erfahrung, ob individuell oder sozial, theoretisch oder praktisch, wurde erst durch die Einfügung in die präferierte Denkweise verständlich. Eine andere und zugleich sinnvolle Auslegungsmöglichkeit innerhalb dieses Rahmens zu finden ist nicht einfach. Soweit das humanistische Konzept historisch der Epoche der Metaphysik und der metaphysischen Auslegung angehört, wie Heidegger kon-

---

3 *GA* 9, 321.

4 Vgl.: „Il faut constater en effet que l'humanisme a été revendiqué au XXe siècle à peu près par toutes les écoles de penseé, matérialistes aussi bien que spiritualistes." Dominique Janicaud, *L'homme va-t-il dépasser l'humain?* (Paris: Bayard, 2002), 17.

5 Die erste Frage Beaufrets an Heidegger lautete: „Comment redonner un sens au mot ‚Humanisme'?" (Auf welche Weise lässt sich dem Wort ‚Humanismus' ein Sinn zurückgeben?) Vgl. *GA* 9, 315, 344.

6 *GA* 9, 321.

statiert, „bleibt jeder Humanismus metaphysisch“,[7] weil dieses Konzept angeblich eine andere Möglichkeit zu fragen weder kenne noch verstehe.

Ein charakteristischer Zug des dominierenden metaphysischen Denkens, dessen Beginn Heidegger mit der Philosophie von Plato und Aristoteles assoziiert, ist die *Seinsvergessenheit.* Die Metaphysik fragt zwar nach dem Seienden in seinem Sein und bestimmt das Sein des Seienden, sie befragt jedoch nicht das Sein selbst.[8] Die Differenz zwischen dem Seienden und dem Sein wird damit nicht erfasst, stattdessen tritt eine Nivellierung beider Begriffe ein. Diese metaphysische Optik, die insbesondere an der Interpretation des vertraulich Bekannten, des relativ klar Vorstellbaren, also an der Auslegung des Seienden orientiert war, wurde vom Humanismus auch bei der Erforschung des menschlichen Wesens übernommen. Auch der Mensch wurde einzig am Bezug zum Seienden ermessen und durch diesen Bezug bestimmt, was die Grundlage für die Ausformung der allgemeinsten Wesensbestimmung des Menschen als *animal rationale* geschaffen hat. Nach Heidegger akzeptieren alle Arten des Humanismus diese Wesensbestimmung als etwas Selbstverständliches, womit sie einerseits auf die fundamentale ontologische Frage verzichten und andererseits den markanten Einfluss der metaphysischen Auslegung auf die Bemühungen des Humanismus, die Menschlichkeit des Menschen zu bestimmen, bestätigen. „Jede Bestimmung des Wesens des Menschen, die schon die Auslegung des Seienden ohne die Frage der Wahrheit des Seins voraussetzt, sei es mit Wissen, sei es ohne Wissen, ist metaphysisch.“[9] Heideggers Humanismuskritik ist deshalb als Kritik desjenigen Menschenbildes zu verstehen, das der Humanismus unter dem Einfluss der metaphysischen Tradition hervorbrachte. Es soll zugleich daran erinnert werden, dass Heidegger dabei nicht auf die Unterschiede der einzelnen historischen Ausprägungen des Humanismus einging, sondern nur die Auslegung des menschlichen Wesens vor Augen hatte, die seiner Meinung nach allen Arten des Humanismus gemeinsam war.

Im *Brief über den ‚Humanismus‘* sind drei Vorwürfe zu finden, die Heidegger gegen die vorherrschende humanistische Auslegung des menschlichen Wesens erhob. Ihm nach beruhe die Definition des Menschen als *animal rationale* auf metaphysischen Voraussetzungen. Der Mensch wird zunächst der Kategorie *animalitas* zugeordnet, die Zuschreibung von *Rationalität* verschafft ihm dann die höchste Stufe in dieser Linie. Dadurch entsteht das Bild des Menschen als des vollkommensten Wesens unter allen Lebewesen, von denen er sich insbesondere aufgrund seiner Rationalität unterscheidet. Heideggers kritischer Ein-

---

7 *GA* 9, 321.

8 *GA* 9, 322.

9 *GA* 9, 321.

wand lautete: Obwohl wir in der Lage sind, auf diese Art und Weise teilweise etwas Wahres über den Menschen auszusagen, müssen wir auch bedenken, „daß der Mensch dadurch endgültig in den Wesensbereich der Animalitas verstoßen bleibt, auch dann, wenn man ihn nicht dem Tier gleichsetzt, sondern ihm eine spezifische Differenz zuspricht".[10] Die Tendenz, das Wesen des Menschen aufgrund der Identifikation seiner biologischen Verwandtschaft mit der Tierwelt zu bestimmen, bedeutet für Heidegger, dass der Mensch nur als ein Seiendes im Bezug auf das andere Seiende ermessen wird, sodass er das Umfeld der Animalität nie verlässt. Die Bemühungen, den menschlichen Organismus naturwissenschaftlich zu erforschen und ihn im Zusammenhang mit der biologischen Evolution zu verstehen, sind für die menschliche Erkenntnis selbstverständlich sehr wichtig. Die Annahme, dass die Erklärung oder Beschreibung des menschlichen Organismus zugleich Erfassung des menschlichen Wesens ist, bleibt jedoch unzureichend. Im Umfeld der *animalitas* kann man die *humanitas* des Menschen, seine Menschlichkeit und Würde nicht entdecken. Nach Heidegger bleiben sie in diesem metaphysischen Entwurf stets übersehen.

Damit hängt auch ein weiterer Punkt der kritischen Reflexion Heideggers zusammen. Es geht um die Anwendung von Regeln der klassischen Definition – die Gattung zu bestimmen und den spezifischen Artunterschied zu finden – im Falle der Begriffsbestimmung des Menschen. Dieses Vorgehens bediente man sich bei der Definition des Seienden, wenn das *Was* des Seienden, also seine Essenz, zu bestimmen und zu benennen war. Nach Heidegger ist diese Haltung, bei der es um Festigung und Stabilisierung der Essenz des Menschen in Form einer Definition geht, der spezifischen Seinsart des Menschen, die sich wesentlich von der Seinsart des Vorhandenseins unterscheidet, nicht gerecht. Bereits in *Sein und Zeit* behauptet er, dass das Wesen des Daseins in seiner Exis-tenz liege, also dass es geradezu aus seinem Sein begriffen werden müsse.[11] Die Existenz interpretiert er dabei nicht als *actualitas*, sondern als Sein, um dessen Begreifen es dem Dasein geht. Im Gegensatz zum metaphysischen Entwurf, wo die Bestimmung der Essenz des Seienden vorherrschte, konzentrierte Heidegger seinen Blick auf die Frage des Seins. Im Bezug zum Sein liegt nach ihm auch der Schlüssel zum Erfassen des Menschen. Wenn dem Menschen die Möglichkeit gegeben ist, sich auf das eigene Sein zu beziehen, es begreifen zu wollen und sich die Frage nach seinem Sinn zu stellen, dann kann er „auch nie als eine spezifische Art unter anderen Arten von Lebewesen gedacht werden".[12] Man kann

---

10 *GA* 9, 323.

11 Martin Heidegger, *Sein und Zeit,* Gesamtausgabe Bd. 2, ed. Friedrich-Wilhelm von Herrmann (Frankfurt am Main: Vittorio Klostermann, 1977), 56.

12 *GA* 9, 324.

eher der Konstatierung von Dirk Mende zustimmen, der sagt: „Heidegger insistiert auf einer *Wesensdifferenz von Mensch und Tier*, die wichtiger sei als die biologische Verwandtschaft. [...] Diese Wesensdifferenz ergibt sich daraus, daß für den Menschen der Bezug zum Sein sein Wesen ausmacht.“[13]

Den dritten Aspekt der Heideggerschen Humanismuskritik bilden humanistische Tendenzen, die zum Anthropozentrismus führen und die mit der Hervorhebung der Einzigartigkeit des Menschen unter allen Tierarten zusammenhängen. Besonders seit der Neuzeit gewann das denkende Subjekt „den Vorrang der Bezugsmitte“.[14] Es wurde bevollmächtigt, Kriterien für die Aufklärung des Seienden zu stellen, es wurde zum „Maß“ bei der Bestimmung seines Seins. Das Seiende galt nur dann als seiend, „sofern es durch den vorstellend-herstellenden Menschen gestellt war“.[15] Das Vor-stellen bedeutete, das Vorhandene als ein Entgegenstehendes vor sich zu bringen und auf sich zu beziehen, also aufgrund von bestimmten Kriterien über die Seinsart des vor uns gestellten Seienden zu entscheiden. Eine solche Herausstellung des Menschen ist nach Heidegger der heikle Punkt des Anthropozentrismus, der den Menschen zur eigenen Verschlossenheit führt. Heidegger hatte deshalb diese Vormachtstellung des Menschen abgeschwächt, wie wir im zweiten Kapitel zeigen werden, womit er insbesondere französische Philosophen zur Kritik am Subjekt anregte.

Wie man sehen kann, hielt Heidegger das Humanismusprojekt aus mehreren Gründen für problematisch, hauptsächlich wegen des starken und einflussreichen metaphysischen Auslegungsrahmens, dem der Humanismus unterworfen war. Heidegger polemisierte dabei jedoch nicht gegen die Idee des Humanismus, sich um die Menschlichkeit des Menschen zu bemühen, sondern gegen die humanistische Interpretation dieser Menschlichkeit und gegen die auf metaphysischen Voraussetzungen basierte Auslegung des menschlichen Wesens. Man könnte befürchten, wie er an mehreren Textstellen betont, dass Polemiken gegen Humanismus zur Befürwortung der In-Humanität und Barbarei führen.[16] In seinem Fall soll es allerdings nicht um die Verteidigung des In-Humanen oder um Herabsetzung der Menschenwürde gehen, sondern um einen Versuch, entgegen der starken Interpretationstradition einen neuen Blickwinkel, eine neue Perspektive des Fragens und Denkens anzubieten, da der metaphysische Humanismus

---

13 Dirk Mende, “Brief über den ‘Humanismus.’ Zu den Metaphern der späten Seinsphilosophie,” in *Heidegger Handbuch. Leben – Werk – Wirkung*, ed. Dieter Thomä (Stuttgart, Weimar: J. B. Metzler, 2003), 249.

14 Martin Heidegger, “Die Zeit des Weltbildes,” in *Holzwege,* Gesamtausgabe Bd. 5, ed. Friedrich-Wilhelm von Herrmann (Frankfurt am Main: Vittorio Klostermann, 1977), 94.

15 *GA* 5, 89.

16 *GA* 9, 330, 345, 348.

„die Humanitas des Menschen nicht hoch genug ansetzt".[17] Damit die Diskussionen über den Sinn und das Wesen der Menschlichkeit nicht nur auf der Ebene der Kritik bleiben, legte Heidegger im *Brief über den ‚Humanismus'* eine existenzial-ontologische Auslegung des menschlichen Wesens vor, die vom Bezug des Menschen zum Sein ausgeht. Dabei geht es um eine Interpretation, die im Hinblick auf sein fundamentales Werk *Sein und Zeit* (1927) zugleich eine Verschiebung in Heideggers Denken bedeutet. Er knüpfte an seine unter dem Titel *Beiträge zur Philosophie (Vom Ereignis)* zusammengefassten Beiträge aus den Jahren 1936 bis 1938 an, die sich mit dem Offenen des Seins aus der Sicht des seinsgeschichtlichen Denkens auseinandersetzen und zugleich ein neues nicht-anthropozentrisches Menschenkonzept entwerfen.

2. Das nicht-anthropozentrische Menschenkonzept

Schon in *Sein und Zeit* deutete Heidegger an, dass der grundlegende Wesens-zug des Menschen in seiner Existenz zu suchen ist.[18] Den Begriff Existenz fasste er dabei ausschließlich als Seinsweise des Daseins auf. Er distanzierte sich von der überlieferten Bedeutung dieses Terminus, wenn existentia im Sinne der Wirklichkeit und das Existieren im Sinne des Vorhandenseins begriffen werden. Die Existenz ist bei Heidegger weder als Wirklichkeit im Unterschied zur Möglichkeit noch als klassische Differenz essentia – existentia, sondern im Sinne des Interesses an eigenem Sein zu denken. Die Sorge um das eigene Sein ist also der charakteristische Wesenszug des Daseins, dem es da-rum geht, das eigene Sein zu verstehen und sich dadurch dem Sein schlechthin zu öffnen. Dank der Sorge ist das Dasein in der Lage, das Besorgen des innerweltlichen Seienden zu überschreiten und in der Berührung mit der Fragilität des Seins (mit der Möglichkeit, nicht zu sein) um das Begreifen des eigenen Seins zu ringen. Diese ständige Sorge um das eigene Sein macht das Wesen des Daseins aus.

Weil es in diesem Konzept noch Spuren der Subjektsouveränität gab, denn die Entbergung des Seins ging immer noch von der Initiative des Daseins aus, unternahm Heidegger in den späteren Texten einen mehr radikalen Schritt in Richtung der Absicht, jedwede mögliche Andeutung des Anthropozentrismus abzuschaffen, womit er der Ablehnung des Anthropozentrismus ein „völlig neues Leben" einhauchte, wie es Janicaud kommentiert.[19] Natürlich war dieser

17 *GA* 9, 330.

18 *GA* 2, 56–57.

19 Vgl.: „Elle a l'immense intérêt – en se voulant plus fidèle que toute autre position à la différence *radicale* qui caractérise l'homme – de redonner une vie complètement neuve au rejet de tout anthropocentrisme." Janicaud, *L'homme va-t-il dépasser l'humain*, 22.

Schritt eng mit dem Versuch um eine neue Interpretation des Seinsbegriffs verbunden. In der späteren Periode setzte Heidegger den um das eigene Sein kämpfenden Menschen in einen breiteren Kontext, er betrachtete den Menschen als Teil des ursprünglicheren und primären Geschehens, das er metaphorisch als Lichtung des Seins oder Wahrheit des Seins bezeichnete. Der Mensch kann das, was er ist, begreifen, „indem er vom Sein angesprochen wird. [...] Das Stehen in der Lichtung des Seins nenne ich die Ek-sistenz des Menschen. Nur dem Menschen eignet diese Art zu sein.“[20] Nach dieser späteren Auslegung bietet sich das Sein dem Menschen an, ungeachtet dessen, ob der Mensch fähig oder unfähig ist, es gleich zu erfahren. Die Tatsache ist, dass unser Sein nicht von uns selbst kommt, sondern dass es uns gegeben bzw. geschenkt wurde. Erst in Berührung mit dem unfassbaren sich gebenden Sein hat der Mensch die Möglichkeit, in sich die Wahrnehmung des Geschehens zu wecken, zu verstehen, dass er ein Teil des umfassenden Ganzen ist, und dadurch sein eigenes Wesen zu erkennen und zum Da-sein, also zur Stätte des Seins zu werden, zur Stätte, wo das Sein sich ereignet. In Berührung mit dem Sein erkennt der Mensch, dass sein Wesen in der Ek-sistenz west. Die Ek-sistenz deutet Heidegger als das „Hin-aus-stehen in die Wahrheit des Seins“.[21] Dieser Interpretation nach kann der Mensch sein eigenes Wesen erst dann verstehen, wenn er aus der Verschlossenheit der eigenen Subjektivität hinausgeht, wenn er seine Ambitionen aufgibt, Zentrum und Mittelpunkt von allem zu sein, und sich schließlich dem umfassenderen, ursprünglicheren Geschehen stellt. Ek-sistieren heißt, aus sich selbst herauszutreten und „in der Nähe des Seins [zu] wohnen“,[22] also sich auf etwas Größeres zu beziehen. Der Mensch hat die Möglichkeit, in sich diesen ekstatischen, exzentrischen Charakter zu entdec-ken und dadurch seine Menschlichkeit zu erkennen. Wir müssen uns jedoch dessen bewusst werden, dass diese Erkenntnis keine Leistung des sich selbst reflektierenden Subjektes ist, sondern die Leistung des Da-seins, das vom Sein angesprochen und damit seiner sich ereignenden Offenheit aus-gestellt wurde. Der Unterschied zwischen der Existenz in *Sein und Zeit* und der späteren Ek-sistenz liegt laut Friedrich-Wilhelm von Herrmann darin, *wie* ekstatisch das Dasein offen ist. Wenn das Dasein ursprünglich vom Sein angesprochen und damit zu seiner Bestimmung geleitet wurde, verwendet Heidegger die Schriftweise Ek-sistenz.[23]

---

20 *GA* 9, 323–324.

21 *GA* 9, 326.

22 *GA* 9, 342.

23 Friedrich-Wilhelm von Herrmann, *Wege ins Ereignis. Zu Heideggers 'Beiträgen zur Philosophie,'* (Frankfurt am Main: Vittorio Klostermann, 1994), 340.

Nach Heidegger wäre es jedoch eine große Verirrung, das Seinsgeschehen theologisch zu deuten. Auch seine Überlegungen aus der späteren Periode, bei denen er versucht, das Sein im Hinblick auf seinen geschichtlichen, sich ereignenden Charakter umzudenken, sind vor allem in der ontologischen Ebene angesiedelt.[24] Das Sein ist hier kein unveränderlicher, letzter, stabiler Horizont, an dem sich alles abspielt, sondern Heidegger versteht es als ein dynamisches „Zwischen", das sich inmitten des wechselseitigen Bezugs des Seins zum Dasein eröffnet. Er denkt es als pulsierendes Geschehen, in dem wir uns das aneignen, was sich uns gibt, wobei sich das Gegebene wiederum uns aneignet. In der Einheit der Aneignung des Da-seins mittels des Seins und der Begründung der Wahrheit des Seins mittels des Da-Seins verwirklicht sich das Wesen des Seins, das er vorläufig als Ereignis bezeichnete.[25] Dieses Ereignis, dessen Grundlage eine Annäherung auf Distanz ist,[26] fasst er jedoch nicht theologisch auf. Obwohl

---

24 Um das Verstehen und die Interpretation des Seins bemühte sich in seinem Denken auch Emmanuel Lévinas. Wenn er das Sein befragt, geschieht das nicht aus einem ontologischen Grund wie bei Heidegger, sondern es geht ihm darum, wie das Sein verlassen oder geschwächt werden kann, damit das Andere (etwas anderes als Sein) zum Ausdruck kommen kann. Róbert Karul untersucht in seiner Monographie *O bytí bez subjektu* [Über das subjektlose Sein] gerade das Lévinassche *il y a* (Es gibt), also das reine Sein ohne Seiendes, wobei er überzeugend und methodologisch fundiert beweist, dass Lévinas in seiner Begriffsbestimmung von *il y a* „nicht das *il y a* selbst, das *il y a* als reines Sein erreicht hat, sondern auf einer ihm vorausgehenden Stufe verharrt". Argumente zugunsten dieser Behauptung sucht Karul in zwei Zugängen: „Während wir beim ersten Vorgehen – durch die Untersuchung der Weltreste im *il y a* – beweisen, dass Lévinas das echte *il y a* nie erreicht hat, werden wir beim zweiten Vorgehen – durch die Untersuchung der Reste des [Lévinasschen] *il y a* in der Welt – die Idee weiterführen, dass es *il y a* überall gibt, dass auch die Bereiche, die Lévinas als Bereiche der Flucht aus dem Sein bestimmt, vom *il y a* bedroht werden und weil sie *bedroht* werden, liegen sie eigentlich im *il y a* selbst. *Il y a* wird es überall geben, wir werden es nicht nur im Genuss, im Wissen, im Wohnen und in der Liebe, sondern auch in der Beziehung ‚face to face' und im Bezug zum Gott finden." Vgl. Róbert Karul, *O bytí bez subjektu. Sproblematizovanie Levinasovho ponímania čistého bytia* (Pusté Úľany: Schola Philosophica, 2012), 49–50.

25 Martin Heidegger, *Beiträge zur Philosophie (Vom Ereignis)*, Gesamtausgabe Bd. 65, ed. Friedrich-Wilhelm von Herrmann (Frankfurt am Main: Vittorio Klostermann, 1989), 262.

26 Eine Dynamik der Grenzüberschreitung vom Sichtbaren zum Unsichtbaren, die Harmonie des Universums und die pulsierende Einheit des Ganzen fasst auch Gaston Bachelard in den Metaphern der Horizontalität und Vertikalität. Eines seiner vertikalen Bilder ist nach Anton Vydra auch der Baum: „Die Vertikalität des Baums, sein Wachstum in zwei Richtungen ist für Bachelards Träumen ein Symbol für das Ganze, das nicht mit dem Auge erfasst werden kann, ‚das so analytisch ist, dass es den Träumenden zwingt, sich einzuschränken'. Der eingeschränkte Träumende wählt immer nur einen Teil des Bau-

er zulässt, dass es auch ein Weg sein könnte, auf dem wenigstens eine Spur der Dimension des Heiligen auftauchen könnte: „Das aber ist die Dimension des Heiligen, die sogar schon als Dimension verschlossen bleibt, wenn nicht das Offene des Seins gelichtet und in seiner Lichtung dem Menschen nahe ist."[27]

Aus dieser anderen Perspektive beruht das Wesen des Menschen auf dem Bezug des Menschen zum ursprünglicherem Sein, laut Heidegger ist der Mensch „insofern mehr denn das *animal rationale*, als er gerade weniger ist im Verhältnis zum Menschen, der sich aus der Subjektivität begreift".[28] Damit der Mensch zur tieferen Erkenntnis kommen kann, muss er auch den Mut haben, etwas einzubüßen. Wenn er sich wagt, auf die relativ starke Stellung eines souveränen Subjekts zu verzichten, kann er in der gewonnenen wesenhaften Armut seine ursprünglichere Bestimmung finden. Einzig dem Menschen ist die Fähigkeit gegeben, sich auf das Sein zu beziehen und es zu begreifen „und nicht nur Natur- und Geschichtshistorien über seine Beschaffenheit und seinen Umtrieb zu berichten".[29] Diese kritische Ablehnung des allgemeinsten humanistischen Menschenbildes führte Heidegger zurück zu erneuten Überlegungen über die Menschlichkeit des Menschen. Deshalb kann man darüber streiten, ob seine Charakteristik des Menschen als Ek-sistenz nicht zugleich eine Art Humanismus darstellt. Er selbst äußert sich dazu nur mit Vorsicht, höchstens könnte es sich seiner Meinung nach um Humanismus einer seltsamen Art handeln. Um Humanismus, bei dem es nicht einzig auf den Menschen ankommt, weil er den Menschen aus der Perspektive der mächtigeren, sich ereignenden Wahrheit des Seins denkt.[30] Er möchte nämlich zeigen, inwieweit der Mensch ein nicht wegdenkbarer Teil des wesenhaften Ereignisses ist. Es wäre also ein Humanismus, der bei der Deutung des Menschen und seines Platzes in der Welt nicht nur die selbstgenügsame menschliche Subjektivität berücksichtigt. Ein Humanismus, der den Schatten des Anthropozentrismus überschreitet. Ungeachtet dessen, ob wir die Bezeichnung Humanismus seltsamer Art auch auf Heideggers Überlegungen über *humanitas* des Menschen anwenden oder darauf verzichten, können wir auf jeden Fall sagen, dass Heidegger mit seiner Auslegung des Menschen als Ek-

---

mes, er sieht den Baum nicht als Ganzes." Deshalb kann man sich nach Vydra dem Ganzen nur im Zustand des Wachtraums annähern. Vgl. Anton Vydra, *Gaston Bachelard a filozofia pohľadu* (Trnava: Filozofická fakulta Trnavskej univerzity, 2012), 97.

27 *GA* 9, 351–352.

28 *GA* 9, 342.

29 *GA* 9, 324.

30 „Das Wesen des Menschen ist für die Wahrheit des Seins wesentlich, so zwar, daß es demzufolge gerade nicht auf den Menschen, lediglich als solchen, ankommt. Wir denken so einen ‚Humanismus' seltsamer Art. Das Wort ergibt einen Titel, der ein ‚lucus a non lucendo' ist." Heidegger, *GA* 9, 345.

sistenz das enge Umfeld der *animalitas* und auch den metaphysischen Auslegungsrahmen, in dem der Mensch lange nur aufgrund seines Bezugs zum Seienden gedacht war, verlässt.

Mehrere Interpreten haben Heidegger vorgeworfen, dass der Mensch viel zu sehr passiv und stark auf die Zuweisung des Seins angewiesen ist, um überhaupt sich selbst verstehen zu können. Sie sehen in diesem Konzept keinen Platz für die Freiheit und Autonomie des Menschen.[31] In der Tat schwächt Heidegger die Stellung des mächtigen Subjekts ab, es wäre jedoch eine zu große Simplifizierung zu behaupten, dass er das Handeln des Menschen in den Händen des anonymen Seins lässt. Einige seiner metaphorischen Wendungen, vielleicht weil es sich eben um Metaphern und Bilder handelt, verführen zwar zum wortwörtlichen Lesen und hinterlassen den Eindruck, dass dem Menschen seine autonomen Kompetenzen abgesprochen sind und dass der Mensch, angewiesen auf die Zuweisungen des Seins, *bloß* „Hirt des Seins"[32] geworden ist. Eine solche Reduzierung war sicherlich nicht Heideggers Absicht, vor allem weil er den unterschiedlichen historischen Ausprägungen des Humanismus vorwarf, dass sie die Menschlichkeit des Menschen zu gering achten. Zugleich erklärt er, dass der eksistierende Mensch im Bezug zum Sein „ein Hörender wird, nicht aber ein Höriger".[33] Wir meinen deshalb, dass es angemessener ist, Heideggers Aussage aus einer anderen Perspektive zu deuten, und zwar als konsequente Absage an Anthropozentrismus. Heideggers kritische Einstellung zur Subjektivität ist aus mehreren seiner späteren Texte[34] bekannt, und sie wird auch im *Brief über den ‚Humanismus'* stringent durchgesetzt. Der Mensch entfernt sich nach Heidegger sich selbst und seiner tieferen Bestimmung, wenn er denkt, das einzige Universum, der Schlüssel zum ganzen Geschehen zu sein. Er sollte lernen, sich selbst als Teil des Seinsgeschehens zu begreifen. Oder, wie es Janicaud kommentiert, der Mensch muss lernen, „sein Sein im Bezug darauf, was sein Kommen auf die Welt veranlasste, und im Bezug auf das Leben, das ihn unterstützt und dessen Träger er ist, zu situieren".[35] Heideggers Ek-sistenz-Konzept will den Menschen nicht seiner Freiheit berauben, es zeigt nur, dass der Mensch erst im Bezug auf

31 Mende, "Brief über den 'Humanismus.' Zu den Metaphern der späten Seinsphilosophie," 256–257.

32 *GA* 9, 342.

33 Martin Heidegger, "Die Frage nach der Technik," in *Vorträge und Aufsätze*, Gesamtausgabe Bd. 7, ed. Friedrich-Wilhelm von Herrmann (Frankfurt am Main: Vittorio Klostermann, 2000), 26.

34 *GA* 5, *GA* 7, *GA* 65 und andere.

35 „Il doit resituer son être par rapport à ce qui l'a fait émerger dans le monde et par rapport à la vie qui le soutient et dont il est porteur." Janicaud, *L'homme va-t-il dépasser l'humain*, 24.

das sich gebende Sein, das ihn wesenhaft ergreift, die Möglichkeit hat, adäquat sich selbst und seinen Platz zu begreifen.

In diesem Sinn betrachtete Heidegger auch Sartres existenzialistischen Humanismus kritisch, obwohl Sartre selbst umsichtig zwei Bedeutungen des Wortes Humanismus unterscheidet. Sartre lehnte die Art des Humanismus ab, die „den Menschen als Zweck und höchsten Wert ansieht",[36] weil er argumentiert, dass der Mensch stets zu schaffen ist und dass wir ihm wegen der bedeutenderen Taten anderer Individuen einen festen Wert nicht zusprechen sollen, um anschließend zu glauben, „es gäbe eine Menschheit, der wir nach Art von Auguste Comte einen Kult weihen könnten."[37] Mit der Ablehnung eines solchen Humanismus wäre auch Heidegger einverstanden. Er polemisierte jedoch gegen Sartres Konzept des sog. existenzialistischen Humanismus, der dem Menschen beibringen will, dass er nichts anderes als sein eigenes Projekt ist. Dass es nichts gibt, was ihn determinieren könnte, dass es keinen Gesetzgeber, „kein anderes Universum als ein menschliches, das Universum der menschlichen Subjektivität"[38] gibt. Gerade dieses einzige Universum, das die menschliche Subjektivität sein soll, bezweifelte Heidegger, indem er sich Gedanken über eine ursprünglichere Dimension, zu der das menschliche Sein gehört, machte. Erst im Bezug auf diese Dimension kann der Mensch sich selbst entwerfen, projektieren, wählen und handeln.[39]

So wie Heidegger die Suche nach der Essenz, nach einer gewissen Natur des Menschen nicht für den Weg hielt, auf dem man zum Begreifen des menschlichen Daseins gelangen könnte, so wie er die Ansicht, dass die Bestimmung des menschlichen Wesens rein Ergebnis der biologischen Evolution ist, nicht teilte, so war er auch mit der Zufriedenstellung und der Verschlossenheit des Universums der menschlichen Subjektivität nicht einverstanden. Seine Auffassung des Menschen als Ek-sistenz ist ein Versuch um eine nicht-anthropozentrische Aus-

---

36 Vgl.: „[...] une théorie qui prend l'homme comme fin et comme valeur supérieure." Jean-Paul Sartre, *L'existentialisme est un humanisme* (Paris: Nagel, 1947), 90.

37 „Et nous ne devons pas croire qu'il y a une humanité à laquelle nous puissions rendre un culte, à la manière d' Auguste Comte." Sartre, *L'existentialisme est un humanisme*, 92.

38 „Il n'y a pas d'autre univers qu'un univers humain, l'univers de la subjectivité humaine." Sartre, *L'existentialisme est un humanisme*, 93.

39 „Eine volle Autonomisierung der Subjektivität gegenüber jeder Alterität" ist mit den Worten von Peter Šajda ein Merkmal, das Theodor Adorno nicht nur in Kierkegaards Subjektivitätskonzept kritisiert, sondern das er auch dem Heideggerschen Dasein aus dem Werk *Sein und Zeit* vorwirft. Diese Kritik bezieht sich jedoch nicht auf das spätere Heideggersche Da-sein. Vgl. Peter Šajda, "Interiorita a indeterminácia: Adornova kritika Kierkegaardovej conceptualizácie vzťahovosti subjektivity," in *Subjektivita/Intersubjektivita*, ed. Róbert Karul et al. (Bratislava: Filozofický ústav SAV, 2011), 37.

legung des Menschen, die mit den naturalistischen Konzepten deshalb konkurriert, weil sie den Menschen aus der Perspektive der Offenheit und des Bezugs zum Sein denkt. Man kann sagen, dass Heideggers Kritik am traditionellen Humanismus den Humanismus vom anthropozentrischen Pathos und von der metaphysischen Optik befreit. Sie mündet jedoch nicht in der In-Humanität, sondern stellt vor das Denken die Aufgabe, den Menschen mehr komplex als nur auf der Ebene seiner biologischen Determinierung zu denken. Uns wird vor allem interessieren, inwieweit Heideggers Polemiken gegen den überlieferten Humanismus und die Diskussionen darüber, was menschlich ist, auch zu den gegenwärtigen Debatten über den sog. Transhumanismus beitragen können.[40]

3. In der Falle des Transhumanismus?

Im breiteren Sinne knüpft die transhumanistische Idee der Verbesserung des Menschen (*human enhancement*) an die überlieferte Idee der humanistischen Veredelung des menschlichen Wesens an, im Unterschied dazu bedient sie sich aber solcher Methoden, die stark von den humanistischen divergieren. Statt mit Hilfe von Bildung und Erziehung versucht der Transhumanismus seine Ziele mit

40 Heideggers kritische Humanismusreflexion, auf die das französische philosophische Umfeld fast sofort reagierte, wurde in Deutschland etwas später von Peter Sloterdijk im Text *Regeln für den Menschenpark* reflektiert. Seinen Beitrag hat er im Juli 1999 auf Schloss Elmau (deshalb oft als *Elmauer Rede* bezeichnet) im Rahmen der Konferenz *Jenseits des Seins – Philosophie nach Heidegger* vorgetragen. Nach Sloterdijk betrat Heidegger bei seinem Fragen nach den Bedingungen des europäischen Humanismus „einen transhumanistischen oder post-humanistischen Denkraum, in dem sich seither ein wesentlicher Teil des philosophischen Nachdenkens über den Menschen bewegt hat". Heideggers Kritik aufgreifend stellt sich Sloterdijk die Fragen: „Was zähmt oder erzieht noch den Menschen, wenn der Humanismus als Schule der Menschenformung scheitert? Was zähmt oder erzieht den Menschen, wenn seine bisherigen Anstrengungen der Selbstzähmung in der Hauptsache zu seiner Machtergreifung über alles Seiende geführt haben?" Dabei antwortet er mit dem Hinweis auf das Potenzial der Gentechnik und auf die Möglichkeiten der negativen und positiven Eugenik, was ihm ermöglicht, kontroverse Gedanken über die Züchtung einer höheren Art des Menschen (in Parallele zu Nietzsches Begriff Übermensch) zu äußern. Der Beitrag zog heftige Reaktionen nach sich, dabei wurde die an die unheilvollen Eugenikpraktiken der Vergangenheit erinnernde Denkweise kritisiert, und es wurden auch Bedenken hinsichtlich der ethischen Folgen der genetischen Züchtung des Menschen geäußert. Die Diskussion brachte jedoch auch Fragen zu der Aufgabe der Philosophie in dieser neuen historischen Situation, in der es zum Wandel des Menschenbildes kommt. Vgl. Peter Sloterdijk, "Regeln für den Menschenpark. Ein Antwortschreiben zu Heideggers Brief über den Humanismus," in *Nicht gerettet. Versuche nach Heidegger* (Frankfurt am Main: Suhrkamp, 2001), 312, 319.

modernen Technologien zu erreichen. Er konzentriert sich dabei auf die biologisch bedingten Mängel und Einschränkungen des Menschen (z. B. Krankheiten, Schmerzen, Leiden, Tod) und sucht nach Methoden, wie die sich schnell entwickelnden Technologien zur Verbesserung der menschlichen Gesundheit beitragen und die physischen, kognitiven oder emotionalen Gegebenheiten des Menschen weitgehend verbessern könnten. Angesichts der Möglichkeiten der transhumanistischen Verbesserung stellt sich wieder die Frage nach der Menschlichkeit des Menschen, diesmal in einer modifizierten Form. Viele fragen sich heute, wie man die Schwelle dessen, was „menschlich" ist, überschreiten, wie man „hinter" die Grenze des Menschlichen geraten beziehungsweise bis zur Kreierung einer neuen posthumanen biologischen Art gelangen kann.[41]

Von mehreren möglichen Arten der Menschenverbesserung, sei es die kybernetische, chemische oder in Form von Technologiekonvergenz, gehört die Modifikation des genetischen Materials zu den meist umstrittenen. Neben Gentransfer innerhalb der somatischen Zellen, bei dem die Veränderungen nur das Individuum betreffen und nicht auf die nächste Generation übertragen werden, erregt viel mehr größeres Interesse und Sorgen die sog. Keimbahnintervention, bei der man aufgrund des Genaustauschs innerhalb der Erbinformation auch den Charakter der nächsten Generationen beeinflussen kann. Auch wenn man zwischen therapeutisch motivierten Eingriffen und Eingriffen, die zur Verbesserung des Menschen führen, unterscheiden kann, liegt zwischen der zur Wiederherstellung des normalen artspezifischen Zustands eines Individuums führenden Therapie und der Verbesserung des Genpools eine hauchdünne und eher fließende als scharfe Grenze.[42] Bei der Verbesserung des Menschen mittels Keimzellenmodifikation besteht nämlich die Möglichkeit, dass wir in den Evolutionsprozess, der bisher ohne menschliche Beteiligung erfolgreich funktionierte, eingreifen und seine Entwicklung infolge der Durchsetzung unserer Ambitionen unwiederbringlich beeinträchtigen.[43] Aber nicht nur das. Strittig ist auch die Frage, ob die zur Entstehung einer posthumanen biologischen Art führenden Veränderungen wirklich zugunsten des Menschen geschehen oder ob der Mensch selbst

---

41 Urban Wiesing, "The History of Medical Enhancement: from Restitutio ad Integrum to Transformatio ad Optimum?" in *Medical Enhancement and Posthumanity*, ed. Bert Gordijn and Ruth Chadwick (Berlin: Springer, 2008).

42 Günther Pöltner, *Grundkurs Medizin-Ethik* (Wien: Facultas, 2002), 160.

43 Diese historische Situation halten viele Autoren wirklich für grundsätzlich. Jürgen Habermas vergleicht sie mit der Kopernikanischen und Darwinschen Wende des geozentrischen und anthropozentrischen Weltbildes. Die dritte Dezentrierung betrifft seiner Meinung nach die Unterwerfung unseres Leibes und Lebens unter die Biotechnik. Vgl. Jürgen Habermas, *Die Zukunft der menschlichen Natur: Auf dem Weg zu einer liberalen Eugenik*? (Frankfurt am Main: Suhrkamp, 2001), 95.

nicht zu einem Instrument wurde, zum Objekt, das je nach dem Willen der vorherigen Generation oder eines gesellschaftlichen Trends vorprogrammierbar und modellierbar ist. Mit einigen positiven Folgen der Verbesserung, wie etwa der Eliminierung von einigen Krankheiten oder der Verlängerung der Lebensspanne, sind auch Elemente der Dehumanisation eng verbunden, wenn die Menschenwürde, die Menschenfreiheit und das Recht auf Chancengleichheit direkt angetastet werden.

Schauen wir uns jedoch dieses Problem aus der Sicht der Fragen an, die für Heidegger und seine Kritik am traditionellen Humanismus ausschlaggebend waren. Heidegger stellte nie die Tatsache in Frage, dass der Mensch auch biologischen Bestimmungen unterliegt, aber er machte darauf aufmerksam, dass darin nicht die ganze Wahrheit über den Menschen west. Seine Kritik am traditionellen Humanismus, der die Größe des Menschen nie genug zu würdigen wusste und den Menschen in das Umfeld des Biologismus einschloss, ist auch für den gegenwärtigen Transhumanismus aktuell. Dieser reduziert nämlich seinen Blick ebenso nur auf die biologischen bzw. genetischen Aspekte des Menschen. Die Verbesserung der genetischen Information, ob mittels Substitution der defekten Gene oder mittels Umprogrammierens des existierenden Codes, ist immer noch vom Biologismus beherrscht. Natürlich können wir bei der Erforschung der Wirkung von Genen auf das menschliche Handeln auch etwas Wesentliches über uns erfahren, aber nach Heidegger ist die Menschlichkeit des Menschen damit noch nicht erfasst. Deshalb kann man sagen, dass vielleicht die größte Gefahr für den Humanismus und Transhumanismus darin besteht, wenn wir uns dieser Reduzierung[44] des Menschen nur auf die Sphäre seiner biologischen Determinierung nicht einmal bewusst sind. So verlieren wir das, was neben der Kraft der Gene unsere Menschlichkeit noch wesentlicher prägt – unseren Bezug zum Ganzen, dessen Teil wir sind, unsere Offenheit zum Sein.

---

44 Zwischen dem Begriff Reduktion, der ein wesentlicher Bestandteil von Husserls phänomenologischer Methode ist, und unserem Ausdruck Reduzierung gibt es einen klaren Bedeutungsunterschied. Die Reduktion bei Husserl fasst Jaroslava Vydrová in ihrer Monographie *Cesty fenomenológie* [Wege der Phänomenologie], in der sie diesen Begriff näher interpretiert, mit folgenden Worten zusammen: „Im Falle der Reduktion geht es um praktische Leistungen, gezielte Schritte – obwohl man ihr manchmal eben Negativität zuschreibt –, wie außer Spiel setzen, Einklammerung, Inhibition als eine Art Verbot, Hindernis, Bremsen, ‚Auflösung des Scheinbaren', Einschränkung." Nach Vydrová liegt das Ziel der Husserlschen Methode der Reduktion darin, dass wir das reine Bewusstsein erreichen und von dort aus unsere Erfahrung deskriptiv beschreiben. Vgl. Jaroslava Vydrová, *Cesty fenomenológie. Fenomenologická metóda neskorého Husserla* (Pusté Úľany: Schola Philosophica, 2010), 20–21.

Im Transhumanismus kam die Idee der Vollkommenheit des Menschen in den Vordergrund, die auch den von Heidegger kritisierten humanistischen Anthropozentrismus beeinflusste. Die extreme Steigerung der menschlichen Vollkommenheit, insbesondere wenn sie mit Hilfe von technischen Mitteln erzielt wird, führte in der Geschichte zu vielen totalitären Projekten. Vielleicht zu den grauenerregendsten gehörten die eugenischen Praktiken der Züchtung einer reinen vollkommenen Rasse. Die transhumanistische Idee, den Menschen bis zur Form einer neuen posthumanen Art zu verbessern, erweckt auch Befürchtungen, dass sie zur Verletzung der Menschenrechte, insbesondere der Gerechtigkeit und der Chancengleichheit führen könnte, da die genetisch nicht verbesserten Individuen im Konkurrenzkampf gegen den verbesserten Teil der Menschheit in eine unvorteilhafte Lage geraten würden. Heideggers Einstellung, solche Ideen einer Polemik und Kritik auszusetzen, ist heutzutage weitgehend berechtigt, damit der Mensch seine Menschlichkeit nicht verliert. Die transhumanistischen Tendenzen stützen sich dabei auf die starke zentrische Selbstwahrnehmung des Menschen, der nach eigener Vollkommenheit und Unsterblichkeit strebt. Eine reale Alternative zu diesen Tendenzen kann das Konzept der exzentrischen, ekstatischen Auffassung des Menschen bilden.

Mit der Humanismuskritik und der Diskussion über Humanismusgrenzen setzte sich Heidegger für die Menschlichkeit des Menschen ein. Denkt jedoch auch der Transhumanismus noch zugunsten des Menschen, wenn er das, was *bloß* menschlich ist, überschreiten will?

4. Schlussfolgerung

Die Entwicklung im Bereich der modernen Technologien eilt heutzutage rasend voran. In einem ähnlichen Tempo verbreiten sich auch die Anwendungsmöglichkeiten für neue technische Verfahren in den unterschiedlichsten Lebensbereichen. Einige von diesen Anwendungen werden reibungslos realisiert, andere dagegen regen leidenschaftliche Diskussionen sowohl in den Fachkreisen als auch in der breiten Öffentlichkeit an. Eines der in der Gegenwart diskutierten Schlüsselthemen ist auch das Thema der transhumanen Verbesserung des Menschen, die mit der Tendenz zur unwiederbringlichen Veränderung unseres bisherigen Menschen- und Weltbildes einhergeht. Ob sie zugunsten der menschlichen Entwicklung und des Lebens an sich geschieht oder nicht, können wir jetzt noch nicht eindeutig beurteilen. Deshalb haben wir in diesem Beitrag, insbesondere in Anlehnung an Heideggers Kritik, gezielt eine Analyse der Beweggründe für Kritik an bisherigen humanistischen Konzeptionen unternommen, um zu erforschen, ob diejenigen Punkte, die sich in den humanistischen Projekten als die strittigsten erwiesen haben und die berechtigt einer Kritik unterworfen wurden, nicht zugleich zum Bestandteil der transhumanistischen Visionen gehören. Wir haben in dem Beitrag gezeigt, dass es mindestens zwei Aspekte gibt, die Heidegger dem Humanismus vorwarf und die man auch in dem gegenwärtigen Transhumanismus identifizieren kann. Im ersten Fall geht es um die Reduzierung unseres Blickwinkels einzig auf den biologischen Aspekt des Menschen, und im zweiten Fall handelt es sich um eine anthropozentrische Auffassung des Menschen, der nach absoluter Vollkommenheit strebt. Beide Fälle sind aus der Sicht des nicht-anthropozentrischen Menschenkonzeptes strittig. Der erste wegen seines reduzierenden Ermessens des Menschen nur im Bezug auf das Seiende, der zweite wegen der Verschlossenheit im selbstgenügenden Universum der menschlichen Subjektivität. Daher halten wir Heideggers Einstellung, dass man immer erneut Fragen stellt und auch alle scheinbar zufriedenstellenden Argumente der Kritik unterwirft, auch heutzutage für eine der Schlüsselaufgaben der Philosophie. Der Philosophie ist es in der Geschichte vielmals gelungen, „auf des Messers Schneide“ zu stehen. In der Gegenwart ist ihr Geschick noch mehr gefragt, damit wir den Ernst des Themas und die Gewichtigkeit der Entscheidung – mit den Worten des Anfangszitats gesagt – „noch rechtzeitig“ erkennen.

Literatur

Bostrom, Nick, and Anders Sandberg. “The Wisdom of Nature: An Evolutionary Heuristic for Human Enhancement.” In *Human Enhancement*, edited by Julian Savulescu and Nick Bostrom, 375–416. Oxford: Oxford University Press, 2008.

Figal, Günter. *Zu Heidegger. Antworten und Fragen*. Frankfurt am Main: Vittorio Klostermann, 2009.

Habermas, Jürgen. *Die Zukunft der menschlichen Natur: Auf dem Weg zu einer liberalen Eugenik?* Frankfurt am Main: Suhrkamp, 2001.

Heidegger, Martin. *Beiträge zur Philosophie (Vom Ereignis)*. Gesamtausgabe Bd. 65, edited by Friedrich-Wilhelm von Herrmann. Frankfurt am Main: Vittorio Klostermann, 1989.

— “Brief über den ‘Humanismus.’” In *Wegmarken*. Gesamtausgabe Bd. 9, edited by Friedrich-Wilhelm von Herrmann, 313–364. Frankfurt am Main: Vittorio Klostermann, 1976.

— “Die Frage nach der Technik.” In *Vorträge und Aufsätze*. Gesamtausgabe Bd. 7, edited by Friedrich-Wilhelm von Herrmann, 5-36. Frankfurt am Main: Vittorio Klostermann, 2000.

— “Die Zeit des Weltbildes.” In *Holzwege*. Gesamtausgabe Bd. 5, edited by Friedrich-Wilhelm von Herrmann, 75–96. Frankfurt am Main: Vittorio Klostermann, 1977.

— *Sein und Zeit*. Gesamtausgabe Bd. 2, edited by Friedrich-Wilhelm von Herrmann. Frankfurt am Main: Vittorio Klostermann, 1977.

Janicaud, Dominique. *L'homme va-t-il dépasser l'humain?* Paris: Bayard, 2002.

Karul, Róbert. *O bytí bez subjektu. Sproblematizovanie Levinasovho ponímania čistého bytia.* Pusté Úľany: Schola Philosophica, 2012.

Kempermann, Gerd. *Neue Zellen braucht der Mensch. Die Stammzellforschung und die Revolution der Medizin*. München: Piper, 2008.

Kristeller, Paul Oskar. *Der italienische Humanismus und seine Bedeutung*. Basel: Helbing & Lichtenhahn, 1969.

Mende, Dirk. "Brief über den 'Humanismus.' Zu den Metaphern der späten Seinsphilosophie." In *Heidegger Handbuch. Leben – Werk – Wirkung*, edited by Dieter Thomä, 247–258. Stuttgart, Weimar: J. B. Metzler, 2003.

Pöltner, Günther. *Grundkurs Medizin-Ethik*. Wien: Facultas, 2002.

Sartre, Jean-Paul. *L'existentialisme est un humanisme*. Paris: Nagel, 1947.

Sloterdijk, Peter. "Regeln für den Menschenpark. Ein Antwortschreiben zu Heideggers Brief über den Humanismus." In *Nicht gerettet. Versuche nach Heidegger*, 302–337. Frankfurt am Main: Suhrkamp, 2001.

Šajda, Peter. "Interiorita a indeterminácia: Adornova kritika Kierkegaardovej konceptualizácie vzťahovosti subjektivity." In *Subjektivita/Intersubjektivita*, edited by Róbert Karul et al., 29–42. Bratislava: Filozofický ústav SAV, 2011.

Von Herrmann, Friedrich-Wilhelm. *Wege ins Ereignis. Zu Heideggers 'Beiträgen zur Philosophie.'* Frankfurt am Main: Vittorio Klostermann, 1994.

Vydra, Anton. *Gaston Bachelard a filozofia pohľadu*. Trnava: Filozofická fakulta Trnavskej univerzity, 2012.

Vydrová, Jaroslava. *Cesty fenomenológie. Fenomenologická metóda neskorého Husserla*. Pusté Úľany: Schola Philosophica, 2010.

Wiesing, Urban. "The History of Medical Enhancement: from Restitutio ad Integrum to Transformatio ad Optimum?" In *Medical Enhancement and Posthumanity*, edited by Bert Gordijn and Ruth Chadwick, 9–24. Berlin: Springer, 2008.

Der Beitrag entstand im Rahmen des geförderten Projektes VEGA Nr. 2/0201/11 „Intentionalität, Affektivität, Existenz in der intersubjektiven Situation" am Lehrstuhl für Philosophie der Philosophischen Fakultät der Universität der Hl. Cyrill und Methodius in Trnava. Ein herzlicher Dank für die Übersetzung des Textes ins Deutsche gehört Herrn Dr. Ján Demčišák, PhD.

—

Jana Tomašovičová
Department of Philosophy
Faculty of Arts
University of Ss. Cyril and Methodius
J. Herdu 2
917 01 Trnava
Slovak Republic
jana.tomasovicova72@gmail.com

Self Shock: The Phenomenon of Personal Non-identity in Inorganic Subjectivity[1]

Corry Shores

Abstract

For both Merleau-Ponty and Deleuze, time is the affecting of self by self.[2] Yet, the nature of this self-affection, as well as the temporal structure it produces, is fundamentally different in each case. In Merleau-Ponty's model, the self's variations are united by time's continuous thrust. However for Deleuze, self variation is far more radical. Our self *immediately* varies from itself: we are always both ourselves and not-ourselves *at every singular moment* and not just over the course of multiple successive nows. While for Merleau-Ponty our temporal moments are organically integrated with each other, for Deleuze time is composed of a series of caesuras which are like the Dedekind cuts that constitute a numerical continuity. We will examine (1) how Merleau-Ponty's organic temporality arises from a subject's continuously integrated personal identity and (2) how in contrast for Deleuze a continuously *disjunctive* temporality is produced by the personal *non-identity* of a *differentially* composed subject. We then evaluate by asking, which model better explains the phenomenon of selfhood, that is, the appearing of oneself to oneself? Building from Deleuze's elaborations and examples, we conclude that we appear to ourselves most phenomenally when we experience differences in ourselves, like when upon looking in a mirror, we become shocked by an abruptly noticed sign of age that makes us for a moment unable to recognize ourselves.

Keywords: Deleuze, Merleau-Ponty, phenomenology, temporal selfhood

1. Introduction

When reading texts by phenomenologists, we find on occasion the authors describing in the first person the phenomenal experiences they are having while writing that given passage. Husserl for example describes the way his brown beer bottle appears to him as the continuous change in daylight alters its tone,[3]

---

1 May I thank Roland Breeur of the Husserl Archives at the University of Leuven, Belgium for contributing his original ideas to this text.

2 Maurice Merleau-Ponty, *Phénoménologie de la perception* (1945; Paris: Gallimard, 2006), 488; *Phenomenology of Perception*, trans. Colin Smith (1962; London/New York: Routledge, 2002), 494; henceforth abbreviated as *PP*, with French/English page numbers. Gilles Deleuze, *Cours 14/03/1978*, trans. Melissa McMahon, www.webdeleuze.com; *Cours 20/01/1981*, www2.univ-paris8.fr, *Cinéma 2: L'image-temps* (Paris: Les éditions de minuit, 1985), 111; *Cinema 2: The Time Image*, transl. by Hugh Tomlinson and Roberta Galeta (Minneapolis: University of Minnesota/Athlone, 1989), 82.

3 "I see a beer bottle that is brown […]. [...] And in this process of appearing there are different appearances," Edmund Husserl, "Seefelder Blättern," in *Zur Phänomenologie des*

and Merleau-Ponty provides a short narration about how the sheets of paper on his desk look differently depending on the light and shadow overlaying them.[4] Because we cannot possibly enter their consciousnesses, these instances instead invite us as readers to direct our awareness to our own immediate phenomenal experiences so that we may follow along using our own personal analyses. So, there is an element of phenomenological analysis that is experiential, and it helps us generate and confirm theoretical claims about the way phenomena appear to us. The phenomenon we will analyze here is our temporal selfhood, which is *who we are* as something self-given through time, and we conduct our analyses by comparing Maurice Merleau-Ponty's and Gilles Deleuze's models for this phenomenon. We will examine the transcendental structures they propose in their accounts, but just as importantly for us, we will also look carefully at the concrete experiences they suggest to help us turn our awareness toward the phenomenal appearing of our own temporal selfhood. This means we would like to know, in which of our experiences do we as temporal selves appear to our self-consciousness? We will find that Merleau-Ponty's model and example direct us to experiences that we have when we notice our self-continuity, that is to say, our having in fact not really changed from how we were in the past. Deleuze's account however calls to our attention moments when we are surprised by how much we have changed. What we will argue is that Deleuze's model is capable of describing a form of temporal selfhood that is more 'phenomenal' in the sense that his differential temporal selfhood stands out and commands our attention much more than Merleau-Ponty's integrated temporal selfhood, and for that reason it is at least as phenomenologically interesting as Merleau-Ponty's model. This is notable both for phenomenologists and for Deleuze scholars, because Deleuze is often considered an anti-phenomenologist, when in fact, as we will see, his model of differential selfhood can make useful contributions to phenomenological studies of time and self.[5] The examples that Merleau-Ponty and Deleuze give are literary, and presumably they intended for us to find in our own experiences similar events in which we became aware of

---

inneren Zeitbewusstseins, *Husserliana X*, ed. H.L. van Breda (1928; Den Haag: Martinus Nijhoff, 1966), 237; "Seefelder Manuscripts on Individuation," in *On the Phenomenology of the Consciousness of Internal Time (1893 – 1917)*, *Edmund Husserl: Collected Works, vol. IV*, ed. Rudolf Bernet, trans. John Barnett Brough (1964; Dordrecht: Kluwer, 1991), 245.

4 "I am sitting in my room, and I look at the sheets of white paper lying about on the table, some in the light shed through the window, others in the shadow," *PP*, 271/262.

5 I describe in more detail the different positions taken in the debate over Deleuze's classification as a phenomenologist or anti-phenomenologist in Corry Shores, "Body and World in Merleau-Ponty and Deleuze," *Studia Phaenomenologica* 12 (2012).

our own temporal selfhood. To further that aim, I will provide my own examples as is common in phenomenological writings, in hopes that readers can better reflect upon their own experiences so to find their own instances as well.

One point of comparison will be the organicism of their models. By organic we mean here a sort of organization in which different parts cohere and work together in formation of a whole that is greater than the sum of its components. In Merleau-Ponty's phenomenology, for example, there is an "organic relationship between subject and world" (*PP*, 189/176); they meld together to form one "flesh,"[6] as when a blind person incorporates a worldly object, a cane, into her body-schema so that she may use her sense of touch to coordinate and navigate throughout her environs (*PP*, 178–179/165–166; 189/176). There is as well in his model an organic integration of our senses: "synaesthetic perception is the rule;" for example we 'feel' the hardness and brittleness of glass when we hear the sound of it shattering (*PP*, 275–276/266–267). In addition, phenomena appear to us through an organic horizonal integration of the parts of our perception: when seeing the red of a carpet, we indirectly have in our awareness the light sources that illuminate it in such a way that it appears with its particular tone (*PP*, 26–27/5).[7] Our focus in the following, however, will be another sort of organic integration in Merleau-Ponty's phenomenology, namely, the synthesis of temporal horizons that is the product of our self-affections. We will examine this sort of integrated temporal selfhood and compare it with Deleuze's alternate model, which is based on the self-affection of a non-temporally integrated self.

2. Merleau-Ponty's Integrated Temporal Horizons

Before proceeding to the topic of temporal selfhood, we will first examine one illustration Merleau-Ponty uses to explain the integration of the temporal horizons of phenomena, namely, the cinematic experiment devised by Russian filmmakers Vsevolod Pudovkin and Lev Kuleshov, which demonstrates the so-called "Kuleshov effect." In this short work, a single image is given and then a close up of a man's face, then another single image, followed again by the man's face, and so on. It is the same footage of the man's expressionless face, but the audience perceived different attitudes depending on the images that preceded it; for example, they thought he appeared pensive when he seemingly

6 Merleau-Ponty, *Le visible et l'invisible,* in *Œuvres*, ed. Claude Lefort (1964; Paris: Gallimard, 2010), 1761–1964; *The Visible and the Invisible*, ed. Claude Lefort, trans. Alphonso Lingis (Evanston: Northwestern University, 1968), 136–138.

7 For my extended analysis of these three sorts of phenomenal integrations in Merleau-Ponty's phenomenology, again see Shores, "Body and World".

looked at a soup bowl but expressed sorrow following an image of a young dead woman: "The audience was amazed at his variety of expression although the same shot had actually been used all three times and was, if anything, remarkably inexpressive."[8]

When the film cuts to the face, the prior image stands nearby on the horizon of our awareness. It does not show itself explicitly, but it appears in the way the man's face seems modified. The apparent hunger in his face attests to the food he seemingly just viewed. Our past intentional acts might have receded from their state of enactment, yet in our horizonal awareness they remain woven throughout the continuous flow of time's passing. This is perhaps why Merleau-Ponty rejects the notion that there can be instantaneous perceptions (*PP*, 26/4); for, all current perceptions are already reaching out to the outer bounds of the present act of awareness, because there is a continuous movement of the organically integrated moments of our phenomenal appearings.

This organic integration holds even for surprising perceptual experiences when the present moment seems uncombinable with the preceding one. In fact for Merleau-Ponty, in these cases each moment is still coherent and integrated with neighboring ones. To illustrate, he has us consider if we were walking along a shoreline until a beached ship comes into view. Not all of the boat is immediately apparent, because we are not yet aware of its masts, which blend in with the trees of the forest behind it. Nonetheless, we eventually become concerned with how some parts of the forest do not seem to fit in with the rest, and then, what previously looked like trees now suddenly appear as ship masts, perhaps because they do not sway in the wind like the trees do. This seems to suggest a discontinuity in the temporal horizons of our awareness. We are retentionally aware of how in the past we saw just trees, all while in this following moment we see some of them as masts. So, this might be a counter-example to the Kuleshov effect, in which our retentions condition the way that we perceive things in our current intentional awareness on account of how they are tightly overlaid and integrated. However, Merleau-Ponty thinks that instead this ship example illustrates the constant, unbreakable integration of phenomenal moments, because even before noticing the masts, we felt implicitly some tension in what we saw, and it was indicative of the new appearance, like how "a storm is imminent in storm clouds" (*PP*, 40–41/20). Although at first we were mistaken, we nonetheless still perceived all the qualities that distinguish the masts from the trees, which suggested to us indirectly that they belonged to the boat rather

---

8 Merleau-Ponty, "Le Cinéma et la nouvelle psychologie," *Les temps moderne* 26 (1947): 937–938; "Film and the New Psychology," in *Sense and Non-Sense*, trans. Hubert L. Dreyfus and Patricia Allen Dreyfus (Evanston: Northwestern University, 1964), 54–55.

than to the forest. Even from the start we had a “vague expectation” that the forest had more to reveal to us:

> The unity of the object is based on the foreshadowing of an imminent order which is about to spring upon us a reply to the questions merely latent in the landscape. It solves a problem set only in the form of a vague feeling of uneasiness, it organizes elements which up to that moment did not belong to the same universe and which, for that reason, [...] could not be associated (*PP*, 41/20).

3. Merleau-Ponty’s Temporal Selfhood

So for Merleau-Ponty, time is constantly flowing, and it is always integrated with itself; yet, our consciousness as well is thoroughly integrated with the flow of time. We ourselves are so much a part of this interlacing of time’s motion that we should not so much think of time as something that flows past and away from us but rather as the motion that springs up continually right here from within us in our immediate experience of it, as if our consciousness itself were the jet of a temporal water fountain.

> We say that there is time as we say that there is a fountain: the water changes while the fountain remains because its form is preserved; the form is preserved because each successive wave takes over the functions of its predecessor: from being the thrusting wave in relation to the one in front of it, it becomes, in its turn and in relation to another, the wave that is pushed; and this is attributable to the fact that, from the source to the fountain jet, the waves are not separate; there is only one thrust, and a single air-lock in the flow would be enough to break up the jet. Hence the justification for the metaphor of the river, not in so far as the river flows, but in so far as it is one with itself (*PP*, 484/489–490).

And our present self, along with our current act of consciousness, are seated at the emerging source of this temporal propulsion: “The passage of one present to the next is not a thing which I conceive, nor do I see it as an onlooker, I effect it; I am already at the impending present as my gesture is already at its goal, I am myself time, a time which ‘abides’ and does not ‘flow’ or ‘change’” (*PP*, 483/489).

To more fully illustrate this integrative flux of phenomenal time, Merleau-Ponty examines a literary example. First consider the consciousness of any character in a novel. More complex characters might have phases in their development. Merleau-Ponty has us recall from Proust’s *Swann’s Way* the character Swann, who falls in love with Odette. Swann experiences both love for her and also jealousy. At first perhaps Odette has love for Swann too. However, she takes other lovers and eventually turns away from Swann, whose unrelenting love causes him to remain attached to her long after her passion for him dies away. Thus his love becomes jealousy. According to Merleau-Ponty, there is an incorrect view that sees his love as coming first and then secondarily it causes his jealousy; this interpretation is faulty, because it only takes into account temporal consciousness as a multiplicity of causally related moments. It does not, however, give us the synthesizing dynamic which is responsible for all moments to be interrelated so thoroughly that each moment indirectly implies the others. It is not so simple that we may say Swann is first in love and secondly jealous: "Swann's love does not cause him to feel jealousy. It is jealousy already, and has been from the start." That coming jealousy tinged his original love; it was on the horizon, implied from the beginning (*PP*, 488/494). Swann's jealous love is also a part of his behavior in general; in other words, even before he fell in love with Odette, his behavior, his manner of integrating with the world, already implied his tendency for this kind of love. It was always on the horizon. Thus, any one part or moment of our consciousness gives us to ourselves as a "comprehensive project" (*PP*, 487–488/493–494). Hence, our consciousness is aware of itself always from a perspective within itself. We cannot think that the self is somehow outside the flow of time while also being aware of itself as being in the flow of time. We cannot, then, regard the self like Kant's transcendental ego. Doing so would mean there is a unified and constant self who is juxtaposed to a stream of self-varying selves given empirically to our senses (and we discuss this Deleuzean "split self" in a forthcoming section). By making this division, we cannot say that the transcendental ego is both the unity of the self lying outside the stream while also always being immanently inside, if in fact we are one and not two distinct selves. So, Merleau-Ponty rejects this view (that Deleuze will instead take up). Thus, the subject must be solely within its temporal flow. Merleau-Ponty regards the self as located directly in the thrusting of time which makes one act of consciousness pass to another. Such a self would explain both the unity of the different acts, as being their glue, and also explain their continuous variation, as being the thrust away from itself.

Hence, each phenomenal moment that ever was or ever will be is thoroughly integrated relationally with all other moments, all organized around the nexus of our immediate phenomenal consciousness. So, our subjective, phenomenal, temporal awareness is at the very basis of the integration of time. Time itself in the way it appears to us phenomenally is virtually indistinguishable from ourselves.

> There is a temporal style of the world, and time remains the same because the past is a former future and a recent present, the present an impending past and a recent future, the future a present and even a past to come; because, that is, each dimension

> of time is treated or aimed at as something other than itself. [...] We are saying that time is someone, or that temporal dimensions, in so far as they perpetually overlap, bear each other out and ever confine themselves to making explicit what was implied in each, being collectively expressive of that one single explosion or thrust which is subjectivity itself. We must understand time as the subject and the subject as time. What is perfectly clear, is that this primordial temporality is not a juxtaposition of external events, since it is the power which holds them together while keeping them apart (*PP*, 484/490).

We will now elaborate this sort of conscious subjectivity which – somewhat paradoxically – is both aware of the flow but also somehow is the actual flowing itself. Consider first how each moment of consciousness is a new one in a succession. But, we are speaking of an act of consciousness that is merely aware of its immediate phenomenal contents. The whole sequence itself (including the now moment) is something else to be conscious of. So that means an additional act of consciousness would be needed in order for us to be conscious of the whole sequence of states that includes the present one in its series. However, that creates another state in the sequence, which requires yet another overarching awareness, and so on to infinity. So this cannot be so. There must be a pre-reflective subjective time-consciousness. Hence, "we are forced to recognize the existence of 'a consciousness having behind it no consciousness to be conscious of it'" (*PP*, 484–485/490–491).[9]

Merleau-Ponty even further elaborates why our consciousness and selfhood are both somehow time itself. We first take into consideration three features of consciousness that will enter into our description. One is that it is a "comprehensive project:" our awareness throughout the time of our life is something whose parts are thoroughly interrelated and integrated with one another. The second consideration is that consciousness is something that appears to itself (*PP*, 487/493),[10] and as we noted, this is a consciousness that is aware both of the entire flow itself all while being inherently a part within it. The third feature is that right now we are *explicitly* aware of certain things, but we are *implicitly* aware of others. This is the case for instance in our vision, where we marginally attend to things in the periphery that are also implied in what we directly see. When viewing a shaded region in the color of a carpet, we are implicitly holding within our peripheral awareness the chair that is blocking the light. And as the Kuleshov/Pudovkin experiment demonstrated, protentions and retentions lie in our marginal temporal awareness but are still expressed implicitly in what appears to us presently.

---

9 Quoting Edmund Husserl, *Vorlesungen zur Phänomenologie des inneren Zeitbewusstseins*.

10 "be apparent to itself" / "s'apparaître".

In our effort to clarify this self-affective dimension, let's briefly note one of Merleau-Ponty's accounts for the way worldly phenomena are given to us. To be consciously aware means to be integrated in a partialized way with the phenomenal world. By this we mean that not everything in the world around us is phenomenally given explicitly all at once. To see the lamp's front side is also to be implicitly aware of how its backside must look from the perspective of the hearth behind it, and in fact, from every other possible vantage-point (*PP*, 96–97/79). Certain implicit aspects, for one reason or another, will call upon us to investigate them, perhaps because something does not seem to fit well with what we currently gather about the situation. Recall the beached boat example, where we were implicitly aware of a few wooden objects as being parts of the ship rather than as being living parts of the forest, although we were not yet explicitly aware of this at first glance. The ambiguity in our awareness motivated us to turn our attention more closely to parts of the forest that before were in the margins of our consciousness. Because there is always something implicit within our explicit awareness calling for our phenomenal investigations, we are continually driven to alter our acts of consciousness. There is thus a perpetual thrusting away from our present act of perceptual awareness. The newer perceptions that we are thrusting toward were previously anticipated acts that we were just protentionally aware of in those prior moments. Our act of awareness of first seeing the masts as trees already had a pull toward the anticipated act of seeing them differently as parts of the boat. Thus, we may more clearly grasp how the temporal thrust expresses itself in our phenomenal relation with the world around us. Consider for example when we try to hold only one thing in our attention for as long as we can, like staring at a still scene or focusing on one concept or image. We eventually feel a tension, a pull away from the focalized act of consciousness toward an awareness looming in the margins of our consciousness.

We emphasize two points from the foregoing analysis. The first is that the temporal thrust is manifest already in the very structure of our present phenomenal awareness, because any given moment of consciousness is structured in such a way that it is already impelled and moving toward a future conscious act that is currently anticipated protentionally. The second point is that it is our consciousness itself that is turning toward its very own implicit horizonal contents. By steering through its own horizons, it modifies the explicitness of contents that were already a part of it. This action is thus both temporal and self-affective.

> Time is 'the affecting of self by self'; what exerts the effect is time as a thrust and a passing towards a future: what is affected is time as an unfolded series of presents: the affecting agent and affected recipient are one, because the thrust of time is nothing but the transition from one present to another. This *ek-stase*, this projection of an

> indivisible power into an outcome which is already present to it, is subjectivity (*PP*, 488/494–495).[11]

This self-modification is also a self-anticipatory movement, because our consciousness knows in advance that it will encounter itself still again as something both continuous with itself yet temporally differential to itself. Our awareness is conscious of both the continuity of itself as a transitional temporal thrust, but as well, it is thereby aware of its self-differential transitional transformations, which are given already in the present not as explicit phenomena but rather as implicit anticipations.

> We must avoid conceiving as real and distinct entities either the indivisible power, or its distinct manifestations; consciousness is neither, it is both; it is the very action of temporalization – of 'flux', as Husserl has it – a self-anticipatory movement, a flow which never leaves itself (*PP*, 487/493).

Our common sense understanding of our selfhood might consider our subjectivity as being a matter of our constant self-identity. Yet, as a transitional thrust, we are instead neither self-constant nor self-identical, even if each moment of our self-displacement is transitionally continuous and horizonally integrated with the rest. For, our temporal thrust is a continuous opening up toward an otherness to our present self. Merleau-Ponty then notes the following observation by Husserl: when we reflect on something in our current awareness, it seems already to be something in the past. Our consciousness, we noted, has a primordial pre-reflective self-awareness. This already-given self-awareness, perhaps, is consciousness' always-already bringing out into explication the implicit parts within it; it would be like a sort of inherent self-curiosity, a constant directedness of our consciousness to those implicit parts in it that most pressingly call out for self-discovery. Our own innate self-consciousness, in other words, could be the engine that propels the thrust of phenomenal temporality. For, just as soon as we turn our explicit awareness to an anticipated protention, we have thereby caused that turning action itself to become an implicit retention, as if consciousness were pushing itself away from itself all while in another sense turning toward itself. So because we are pre-reflectively self-aware consciousnesses, we have by default a thrusting flux of temporal awareness.

> The fact that even our purest reflection appears to us as retrospective in time, and that our reflection on the flux is actually inserted into the flux, shows that the most precise consciousness of which we are capable is always, as it were, affected by it-

11 Citing Immanuel Kant and also Martin Heidegger, *Kant und das Problem der Metaphysik.*

> self or given to itself, and that the word consciousness has no meaning independently of this duality (*PP*, 489/495–496).

We may finally summarize all the combined features of our consciousness' temporal thrust. *The thrust is our personal subjectivity*, because its continuous transitional action is what makes all our other acts belong integratively with each other and also belong continuously to our current act of immediate awareness. *The thrust is temporal*, because its movement displaces acts of awareness out of present consciousness into implicitly retended awarenesses of past acts, and it brings out from our marginal awareness implicit phenomenal contents that were previously anticipated protentionally as future acts. *The thrust is self-aware*, because it is the ongoing motion of one part of our consciousness, the explicit part, bringing into its awareness another part of our consciousness, the implicit part. And thus *the thrust is self-affective*, because consciousness is this transitional motion that changes its own parts either from implicit to explicit modifications or from explicit to implicit modifications. Then finally, *the thrust is immediately experiential and phenomenal*, because the implicit and explicit phenomena result from our given interrelation with our entire phenomenal world at some present moment in time.

> The primary flow, says Husserl, does not confine itself to being; it must necessarily provide itself with a 'manifestation of itself' (*Selbsterscheinung*), without our needing to place behind it a second flow which is conscious of it. It 'constitutes itself as a phenomenon within itself'. It is of the essence of time to be not only actual time, or time which flows, but also time which is aware of itself, for the explosion or dehiscence of the present towards a future is the archetype of the *relationship of self to self* (*PP*, 488–489/495).[12]

As we will see, Deleuze also regards phenomenal time as being a matter of transition or passage, however its temporal parts do not organically integrate but are rather like a series of cuts in time. For the sake of our comparison, let's briefly attend once again to Merleau-Ponty's literary illustration from Proust's *Swann's Way*. Merleau-Ponty's point is that Swann's subsequent jealousy was already a part of his initial love; it was from the beginning a jealous love. We wonder then, how might we characterize the way that Swann's temporal selfhood appeared to him when the jealousy of his love became more evident? It would seem to be the realization that he was jealous all along, even in the apparent purity of his prior state of love. This is similar then to Merleau-Ponty's example of the ship masts, where initially we are implicitly aware in our protentions that later the forest will appear differently. Similarly, Swann originally was proten-

---

12 Citing Husserl, *Vorlesungen zur Phänomenologie des inneren Zeitbewusstseins* and Heidegger, *Kant und das Problem der Metaphysik*.

tionally aware of his forthcoming jealousy, although he was only implicitly conscious of this at the time. So, when we notice something about ourselves that we previously protended implicitly, we realize that in fact we have not changed; it is an appearing of our self as having always been the same self, like how Swann might realize he is always a jealous lover. Yet, consider if we look in the mirror, and somehow it comes to our attention that we have not changed. How could this be an awareness of the temporal nature of our selfhood? It would seem instead to indicate that we have in some way been untouched by time. We turn then to Deleuze's model and illustrations to see if they are more able to bring to our awareness the temporal significance of our selfhood.

4. Deleuze's Temporal Synthesis

Deleuze's exposition of self-affected temporality is found largely in what he calls the "third synthesis of time." The first synthesis is the living present that fuses *now* with *now* with *now*, etc. The second one, memory, places *nows* into relations of *now* and *prior*, while the third synthesis relates *now* to *next*, and it is the "empty form of time" bearing the structure of the split self as well as being the basic temporal relation of *before* and *after*. Yet, first we should briefly examine the basis for the time syntheses that Deleuze provides in *Difference and Repetition*.

Deleuze begins his exposition of the syntheses with Leibniz' concept of *mens momentanea*.[13] At the moment two bodies collide, they leave upon one another impressions of their motional tendencies, but they last only as long as the transfer, which is instantaneous. So, physical bodies without memories would be momentary minds (*mens momentanea*), because none of the impressions that they receive will last for longer than an instant.[14] This means also that for such bodies there is no such thing as time. If all you have is *now*, then you do not have time, because time requires temporal differentiation: it needs both a *before* and *after*, a temporal ordering, and not merely a now. Yet, this exclusive now of momentary minds is not the same as saying there is just one now-moment that is suspended forever, because on the contrary, nothing stays exactly the same for more than an instant. So for momentary minds, there is just the

13 Deleuze, *Différence et répétition* (Paris: Presses universitaires de France, 1968), 96; *Difference and Repetition*, trans. Paul Patton (New York: Athlone, 1994), 70; henceforth abbreviated as *DR*, with French/English page numbers.

14 Gottfried Leibniz, "Theoriae motus abstracti Definitiones," in *Die Philosophische Schriften, vol. 4*, ed. C.I. Gerhardt (Hildesheim/New York: Geog Oms, 1978), 229–230; "The Theory of Abstract Motion: Fundamental Principles," in *Philosophical Papers and Letters*, ed. and trans. Leroy E. Loemker (1956; Dordrecht: Kluwer, 1989), 140–141.

given present now, which bears no temporal relations to other nows. For, the other nows cannot exist for things whose impressions vanish with each successive moment. This also means that there is no basis for momentary minds to have subjectivity, because this requires temporal self-differentiation, as we will soon see.

Unlike such simple physical bodies, we are beings with memories. Consider the most basic sort of non-momentary mind. The simplest cellular microorganisms are themselves composed of momentary minds; for example, a cell might be made up of such elements as water, nitrogen, carbon, and so on (*DR*, 102/75). And yet, the whole conglomeration acts survivalistically in its current now moment, which means it anticipates being a continuation of itself after this now moment ceases. It has some kind of an implicit knowledge that it has a past and a future. Deleuze says that the genetic heritage of cellular organisms is their awareness of the past, and their current needs are their awareness of the future (*DR*, 100/73). This is perhaps the most primitive form of self-consciousness possible. In fact, this model of the single-cellular organism will help us grasp the *a priori* synthetic unity of our own self-consciousness. Kant will emphasize that transcendental self-consciousness is not self-cognition. So, we are not saying that the cell is thinking about itself. However, it regards itself as a unity existing beyond the present, despite the fact that nothing physically exists beyond the given now, and this is evidenced by the fact that right now it is seeking sustenance that will allow it to survive beyond the present moment.

We, however, have a far more sophisticated perceptual and cognitive apparatus than mere single-celled organisms have. So for us, this basic self-consciousness can take on a more complex and amplified form. We can appear to ourselves; we can mentally reflect on ourselves with our imagination; we can conceptually understand ourselves as well too. But cognizing ourselves, Kant reminds us, requires not just concepts but also intuitions.[15] So, we need to appear to ourselves in order for us to have self-recognition. However, our self-appearances are never the same, because the one that is given now is just a fleeting presentation of ourselves. We notice, then, that there is a tension already between the trans-temporal unity of self-consciousness and the complete instability of the now in which self-consciousness appears to itself; for, the now remains self-same for no more than a fleeting instant. And for Deleuze, it is this tension that creates the phenomenon of time. We say *phenomenon*, because un-

15 Immanuel Kant, *Kritik der reinen Vernunft, Erster Teil, Werke Vol. 3*, ed. Wilhelm Weischedel (Darmstadt: Wissenschaftliche Buchgesellschaft, 1968), A51/B75; B154–156; *Critique of Pure Reason*, trans. Paul Guyer and Allen W. Wood (Cambridge: Cambridge University, 1998); henceforth abbreviated as *KrV*.

der this framework of *mens momentanea*, time does not exist in actuality. Time can only become manifest when it is presented as an appearance given to beings that are able to distinguish *before* from *after*. So, there is a self-consciousness now that assumes itself to exist in moments beyond the now, even though nothing physically exists beyond the now. Thus, this self-consciousness assumes itself to manifest in different nows, which means it takes a manifold variety of itself and regards them as all belonging to itself, despite the impossibility of a now moment entering into any sort of actual relationship with the other non-existing nows.

To help us grasp what it is like for our self-consciousness to recognize itself, Deleuze evokes two ancient Greek mythological figures, Narcissus and Actaeon. We might at first think that self-consciousness is like Narcissus' fixed gaze on his own image on the pool's surface. But Deleuze explains that Narcissus only illustrates the pleasure we seek in self-recognition. Our experience of self-appearing is in fact more like Actaeon's transformation (*DR*, 102/74–75). After he happened upon goddess Diana bathing naked in the woods, she turned him into a deer. He becomes aware of this when seeing his new animal form reflected in the water.

> This said, the man begun to disappear
> By slow degrees, and ended in a deer.
> A rising horn on either brow he wears,
> And stretches out his neck, and pricks his ears;
> […]
> But as by chance, within a neighb'ring brook,
> He saw his branching horns and alter'd look.
> Wretched Actaeon! in a doleful tone
> He try'd to speak, but only gave a groan.[16]

Actaeon groans because he both recognizes it is himself, for it is his self staring back at him, while at the same time, he must recognize himself as becoming something completely different than who he had been right up until then. In a manner of speaking, to be self-conscious means we must look in the water and see a deer, and yet still identify with that other, as if to say, "that is me, but who am I?" Being a self means being an other to oneself. This is the basic model for Deleuze's phenomena of conscious temporality and selfhood. So, every now involves the destruction of some other now, but self-consciousness allows non-existing nows to somehow coexist incompatibly with existing ones.

16 Ovid, "The Transformation of Actaeon into a Stag," in *Metamorphoses*, trans. Joseph Addison (Ware: Wordsworth, 1998), 78.

Hence, the unity of self-consciousness already implies temporality, but this is not time as a flow of succession, but rather time as a form, as an *a priori* synthesis of incompatible states of affairs. So self-consciousness, as it appears to itself right now, is regarding itself as already formally linked to what it becomes, even though it will be an incompatible variation of itself. Thus we, through our unity of self-consciousness, are already cutting ourself off from ourself. There is a cut between Actaeon the man and Actaeon the stag. But Actaeon, even before actually seeing his reflection, while just first beginning to peer into the water, already regards whatever he is about to see as being an appearance of himself. Our already regarding ourself as united with our forthcoming differential self-appearing is not merely this cutting into two incompatible variations of ourself; it is also the formal production of time, because it creates the phenomenon of temporal difference of *before* and *after*, of *now* and *next*, when otherwise to the momentary minds of the world there is just the *now*, *now*, *now*, *now*… As we will continually see, the phenomena of selfhood and temporality are fundamentally co-given appearances. This is the third synthesis of time. It is the cut that produces the bare order of time by splitting *the now* into *the now and the next*. It brings *the next* into a differential relation with *the now*, and thereby makes next and now be given immediately with each other, at least in a formal relation.

5. The Third Synthesis of Time

We will focus our examination of the third synthesis on Deleuze's notions of *time is out of joint* and the split self; for as we shall see, they involve time itself as a phenomenon that appears to our explicit awareness. From the beginning to the very end of Deleuze's philosophical writing career, he rehearsed his "time is out of joint" argument at least nine times, in each case with slight nuances of emphasis and elaboration: in *Difference and Repetition*, in the introduction to his Kant book, in an extended version of that introduction published separately, in at least five of his class sessions from 1978 to 1984, and in his coauthored work, *What is Philosophy?*[17]

17 Deleuze's rehearsed argument can be found in: *DR*, 116–123/85–91; *Kant's Critical Philosophy: The Doctrine of the Faculties*, trans. Hugh Tomlinson and Barbara Habberjam (London: Athlone, 1995), vii-ix; *Critique et Clinique* (Paris: Les éditions de minuit, 1993), 40–45; *Essays Critical and Clinical*, trans. Daniel W. Smith and Michael A. Greco (Minneapolis: University of Minnesota Press, 1997), 27–31; *Cours 14/03/1978*; *Cours 21/03/1978*, trans. Melissa McMahon, www.webdeleuze.com; partly in *Cours 12/04/1983*, www.webdeleuze.com, www2.univ-paris8.fr; *Cours 13/12/1983*, www.webdeleuze.com, www2.univ-paris8.fr; *Cours 17/04/1984*, www2.univ-paris8.fr; and Gilles Deleuze and Félix Guattari, *Qu'est-ce que la philosophie?* (Paris: Les éditions

Deleuze's account begins with his careful analysis of Kant's critical revision of Descartes' cogito formulation. Mere consciousness of itself is apperception, and this is the simple representation of the I (*KrV*, B68). The self is split because we have two different and unbridgeable self-consciousnesses, the transcendental and the empirical apperception. But here we should first describe briefly Kant's three time syntheses: apprehension, recollection, and recognition. We (1.) apprehend representations given in intuition one instant to the next. The reproductive imagination (2.) gathers the passing apprehensions and the productive imagination synthesizes them together in such a way that the understanding (3.) may recognize the synthesized intuitive manifold by means of its concepts (*KrV*, A98–A110; B141; B150–152). For example, when viewing a triangle, we apprehend the parts of its sides little by little. The reproductive imagination retains these tiny and otherwise fleeting apprehensions while the productive imagination reconstructs them into an image of the triangle. It performs this reconstruction in accordance with our pure geometrical concept for triangles, by employing a general procedure or rule (a "schematism") that is used for forming images of such pure shapes in space (*KrV*, A137–142/B176–181). Yet, note the tension between the temporal flux and the conceptual unification. The moments of apprehension, although temporally exclusive to one another, are taken together simultaneously in the reproductive imagination. There must somehow be a unified consciousness that is the same throughout the moments, for otherwise they would belong to a different consciousness and thus be unavailable for synthesis. This *a priori* unified self-consciousness is the *transcendental apperception*, the "I think," which accompanies all of our conceptual acts. The *empirical apperception* is our self-consciousness as under a stream of alterations (*KrV*, A107). We are phenomenally aware of ourselves as ever inconsistent and disunified. So, empirical apperception is not enough to allow us to apprehend ourselves as enduring persons, because it has no means to make all our variations belong together. Yet, transcendental apperception has no intuitive content, so it cannot alone apprehend itself or cognize itself. In order to do this, it needs to synthesize itself into a manifold of the phenomenal self-appearings happening at temporally distinct moments. In other words, the consciousness of our unified non-temporal subjectivity, in order to become explicitly aware of itself, needs to manifest itself as being disunified through time.

---

de minuit, 1991), 34–36; *What Is Philosophy?* trans. Hugh Tomlinson and Graham Burchell (New York: Columbia University Press, 1994), 31–32. Given the consistency between the accounts, we will draw from all of them without citing any in particular, although we will make note when specific texts are exclusive or notable sources for the material in question.

The way transcendental apperception appears phenomenally is by means of self-affection. External objects modify our outer sense, but self-affections modify our inner sense (*KrV,* B156). Kant offers an example of self-affection: our acts of attention (*KrV*, B156–157). When we focus our attention, we notice a modification of our awareness, its becoming more focused. This resulted not from an external object, but from an internal influence. It gives us a representation of our active subjectivity as a self-affecting subject, but it appears only phenomenally as a passively modified empirical self-consciousness.

With these distinctions we may better examine Deleuze's analysis of Kant's version of "I think, therefore I am." Descartes assumed that the act of self-consciousness was enough to determine oneself as a thinking being; and, by knowing that one is a being of some determinate sort, one would then know oneself to exist. But as we noted, Kant's transcendental apperception, the "I think," cannot cognize itself as a thinking being without manifesting itself to the empirical consciousness. So Kant's "I think" alone cannot determine its existence as a thinking being.

> In the synthetic original unity of apperception, I am conscious of myself not as I appear to myself, nor as I am in myself, but only that I am. […] Now since for the cognition of ourselves […] a determinate sort of intuition […] is also required, my own existence is not indeed appearance […], but the determination of my existence can only occur in correspondence with the form of inner sense, […] and I therefore have no cognition of myself as I am, but only as I appear to myself. […] I exist as an intelligence that is merely conscious of its faculty for combination (*KrV*, B157–158).

To clarify the matter, Deleuze draws the following four distinctions from Kant's formulation: the undetermined, the determined, the determination, and the determinable. The *undetermined* here is our existence, the "I am." It is what will undergo the determination. The *determined* (the result of the determination) is that "I am a thing that thinks." The *determination* (what is doing the determining) is the "I think." Yet the "I think" is self-consciousness and is not self-cognizant, so it alone is not enough to determine its own existence. As we noted, it can only determine itself by synthesizing a manifold of its own appearings, which means it must appear in self-disunity over successive moments of time. So, Deleuze says that Kant introduces a fourth element, determinability.[18] The act that will determine our existence, the "I think," can only determine our existence as a being who thinks if we appear in intuition, which for Kant means on the basis of the *a priori* pure form of time (*KrV*, A30–41/B46–58). Our *determi-*

18 We draw these four distinctions from a number of the nine aforementioned sources, but especially from the extended elaboration in Deleuze, *Cours 17/04/1984*.

nability, then, is our synthesis between, on the one hand, our empirically and temporally varied self-appearings with, on the other hand, our atemporal unified intellectual self-consciousness. Pure temporal relationality, then, is our determinability: it is the tension between self-unity and self-disunity, making moments distinct while also forcing them together. It is the fundamental contraction of our self-consciousness' two incompatible tendencies toward temporal disintegration and unified self-identity. So, the transcendental apperception is our personal identity in the sense that it is consciousness of itself as self-same. The empirical apperception is our non-personal non-identity, because the different self-appearings do not by themselves belong together to a self-identical consciousness. Our determinability, however, is our *personal non-identity*. It is personal, because all our temporally distinct phenomenal variations belong together through the action of the synthesis. However, it is our non-self–identification, because our unified self-consciousness is represented in our empirical self-consciousness but is so transformed by temporality that it cannot be identified with itself anymore.

Hence, Deleuze turns to Rimbaud's formulation "Je est une autre." He uses it in two letters from around the same time:

> I is someone else. It is too bad for the wood which finds itself a violin.
> For *I* is someone else. If brass wakes up a trumpet, it is not its fault.[19]

For Rimbaud, Deleuze explains, these parings are based on concept-object relations, where the concepts *trumpet* and *violin* actively give form to the wood and brass, which are shapeable substrates; and thus Rimbaud thinks he likewise can become molded into a poet. What makes Kant's "I is an other" different is that the I is not a concept for ourselves but is rather the representation that accompanies all our concepts; and the self is not an object being formed by concepts, but is rather the continuous variation of our successive states and the "infinite modulation of its degrees at each instant." The I-Self relation is a modulation and not a mold, because we are not being shaped by forms but are rather continuously breaking out from forms.[20]

Yet, the third synthesis is not just a crack in the self, it is also a "cut" in time, a *caesura*. To elaborate, Deleuze turns to Hölderlin's analysis of Sophocles' Oedipus trilogy. But first we need to examine cosmic cycles and the "cardinal" measuring of time by means of circular returns of motion. If we could watch the stars move for 24 hours, we would see them make a complete circular

---

19 Arthur Rimbaud, "Letter To George Izambard, 13 May 1871" and "Letter To Paul Demeny, 15 May 1871," in *Complete Works, Selected Letters: A Bilingual Edition*, trans. Wallace Fowlie (Chicago: University of Chicago, 1966), 304–305.

20 Deleuze, *Critique et Clinique*, 44/30.

rotation around the North or South Star, depending on our location. This is because the earth makes daily rotations on its axis, which points to these polar stars that remain fixed in their location in the sky all while the others circulate around them. But because the earth moves little by little around the sun, each night any particular star will appear in the same position at a slightly different time. This means that if we were to observe a specific star's location every day at the same time, for instance if we were to note where a star is at midnight every night of the year, we would find it also makes a yearly circular rotation around its polar star (figure 1).

Fig. 1: Circular path of the stars (by LGCS Russ)[21]

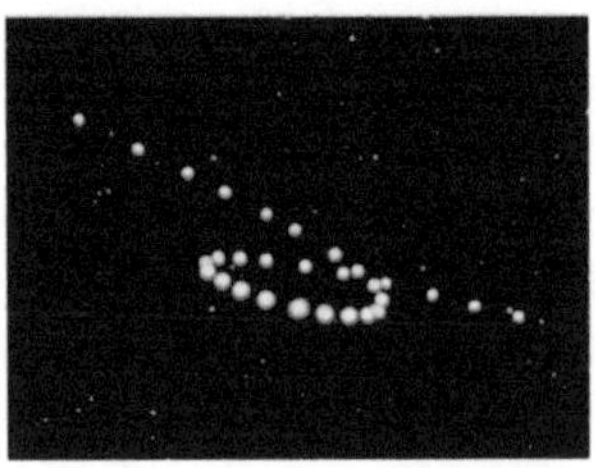

Fig. 2 Deviant movement of Mars (© Tunç Tezel)[22]

Now, the sun's or any other star's quick return to a designated starting point in the sky marks a day. It gives us one unit of time against which to measure other times. And so the yearly circling of the stars around a polar star is meas-

21 LGCS Russ, "Circumpolar AZ81," http://commons.wikimedia.org/wiki/File:Circumpolar_AZ81.jpg

22 Tunç Tezel, "Retrograde Mars," © Tunç Tezel, all rights reserved, http://apod.nasa.gov/apod/ap031216.html

ured as being about 365 daily sun circuits, just as the day is measured as being around 1/365th of a yearly star cycle. Likewise, the circulation of the hour hand in a clock returns to its starting position 24 times a day, the minute hand 60 times per every hour-hand rotation, and so on. Thus, regular cycles of circular motion provide the standard units for measuring time. This might surprise us, because time measured as being divisible into parts that are external to one another – like the individual hours of a day – is extensive time, which we normally characterize as a line, as for example when we draw out a time-line of events. And it is "cardinal" time, in the sense of the cardinal numbers 1, 2, 3, etc., because the cycles are distinguished from one another as independent, enumerated entities. Deleuze, however, is claiming that in fact extensive cardinal time moves in a circle, while there is also an intensive pure form of time that breaks out of circular rotation into the form of line, as we are going to see. This is "ordinal" time, like the ordinals 1st, 2nd, 3rd, etc., because it does not so much enumerate successive moments as much as it designates one moment as coming before or after another one.

Now, consider how the planets are not fixed points of light out in the distant cosmos like the stars are; rather, they orbit the sun like the earth. So in the sky, they do not follow the stars' same yearly circular paths around a polar star, but instead make their own motions that go against the grain of the stellar yearly circulations. In fact, they might at times change direction and move backwards when in retrograde motion in the earth's sky (figure 2, above). Yet still, their circulations in our night sky eventually bring them back to some starting point. When discussing the temporality of Greek tragedy, Deleuze has us regard the stars regular circuits to be like the standards or limitations the gods place on men. Mortals should not stray from the fixed divine laws. Yet in tragic stories, characters will break some law or limitation, transgressing that limit. So, we are to think of the planets moving against the stars' circular paths as being like a tragic character having committed some injustice, taking him or her on a deviant path in defiance of divine law. Yet, after a long while, the planets do return to a given starting point. In a way, this is like justice eventually being reinstated. However, in these ancient tragedies, the wrong-doings are normally rectified through vengeance; some other character murders the current guilty one, and that new character now restarts the same tragic cycle, only for it to be continually repeated over and over again.[23] Now, as we will see, Deleuze's account of Oedipus' shocking confrontation with himself as being someone contrary to the person he rightfully considered himself to be will provide us with a template for finding in our own experiences those moments when our differen-

23 These astronomical elaborations are found especially in Deleuze, *Cours 21/03/1978.*

tial temporal selfhood is given directly to our focal awareness. Thus because of its central role in our studies here, we will need to carefully work through Deleuze's elaborate commentaries on the story.

Let's then first notice the cyclical patterns of vengeance in Aeschylus' tragedies before seeing how they are broken in Sophocles' Oedipus trilogy. Clytemnestra, for example, plots with her lover, Aegisthus, to murder her husband Agamemnon. Clytemnestra wants justice for Agamemnon having killed their daughter when he needed to appease the gods. Agamemnon reluctantly did this in order to obtain the winds needed to carry the Greek fleets home after their victory over Troy. And, Aegisthus wants to avenge a crime that Agamemnon's father committed against his own father. So, Agamemnon transgressed the limits with his murder of their daughter. When Clytemnestra kills Agamemnon in return, that brings justice, but is itself a new transgression. In the next part of the *Oresteia*, Orestes, son of Clytemnestra and Agamemnon, returns after some years in order to murder his mother to avenge his father. Thus, this sort of temporality in a way has fixed points that serve to orient the phases of the movement, something like how the cardinal directions designate unique points of reference. And, just like how the planets rotate around the sun, the circular motion of this time seems to turn round and round as if on a hinge (*cardo*).

Deleuze then turns to Hölderlin's analysis of Sophocles' Oedipus trilogy to introduce the notion of the temporal cut or caesura. Recall the story of *Oedipus Rex*. Thebes is cursed with a plague. This is because its king Oedipus has committed the crimes of killing his father and marrying his mother. Yet despite prophesies foretelling this, Oedipus and his mother Jocasta are unaware. Long in the past, Oedipus' father Laius heard the prophecy that his son would kill him. So he arranged for his newborn son Oedipus to be left on a mountain to die. Instead, the child was saved and adopted as the prince of Corinth. Oedipus later wanted to know who his true parents were. The oracle told him of his curse, that he would kill his father and marry his mother. While traveling to Thebes, he fights and kills a man who was really Laius, his father, but he did not know it then. After solving the sphinx's riddle, Thebes makes Oedipus king, and he marries Jocasta, who is his mother, but not yet to their knowledge. Creon, Jocasta's brother, was sent to Delphi to find out how to lift the plague over Thebes. He learns that they must cast out Laius' killer. Tiresias the seer says it is Oedipus. Clues slowly emerge that reveal this to be true, and Oedipus finally recognizes his crimes. Jocasta realizes too, and she hangs herself. Oedipus we know blinds himself by gouging out his own eyes. He entrusts his daughters Antigone and Ismene to Creon, and he is exiled from Thebes.

The Sophoclean tragedies, Hölderlin explains, have an onrush of change, culminating at a highest point. To counter-balance the disproportion in the pace of the drama, what is needed at that highest point is a "counter-rhythmic rupture," which he calls the *caesura*. So in the rhythmic sequence of a tragic play, there will be two main parts that are divided by the caesura. Both portions will relate to each other in a way that makes them feel as though they have common weight, even though they move at different speeds. When the initial part of the tragedy has more 'pressure', the counter-rhythmic caesura will be placed more toward the beginning. When instead the ending part has more pressure, the caesura will be placed toward the end. This is because the more rapid parts seem to weigh more. *Oedipus Rex*'s caesura is found toward the beginning when Tiresias tells Oedipus he is Laius' killer.[24] Because this fact is spoken by a prophet, it has the weight of the truth, despite Oedipus already being completely justified in concluding it must be false. So at the moment of the prophesy, Oedipus on the one hand still has reason to affirm the state of affairs "I am not my father's killer" while now having other reason to affirm that in fact "I am my father's killer." At the point of rhythmic rupture at the caesura, he must accept both states of affairs even though they are incompatible.

So recall the split self with regard to Kantian self-consciousness. There is a self-consciousness that is *a priori* unified, and on the basis of its unity, it may synthesize its empirical appearings despite them being temporally incompatible. However, that means we do not cognize ourselves as we are in our self-same identity, but rather insofar as we appear through a manifold of self-variation in time. In the case of Oedipus, he in one moment affirms "I am not Jocasta's son" while just after that must affirm "I am Jocasta's son." In both cases it is the same I, but the predications push against each other. His empirical self-consciousness does not identify the Oedipus in one moment with the Oedipus in the next moment, because they are temporally exclusive to one another; yet his unified *a priori* self-consciousness forces the "I" in both cases to identify and thereby combine his incompatible moments of empirical self-consciousness. Oedipus' being, then, is not so much a matter of what makes him always self-same; instead, it has more to do with what allows him to always be posed in opposition to himself. As Hölderlin explains:

> Being must not be confused with identity. If I say: I am I, the subject ("I") and the object ("I") are not united [...]; on the contrary, the I is only possible by means of this separation of the I from the I. [...] How can I say: "I"! without self-consciousness? Yet how is self-consciousness possible? In opposing myself to myself, separating myself from myself, yet in recognizing myself as the same in the op-

---

24 Friedrich Hölderlin, *Sämtliche Werke, vol. 5* (Stuttgart: W. Kohlhammer, 1952), 195–197; *Essays and Letters on Theory*, ed. and trans. Thomas Pfau (Albany: SUNY, 1988), 101–102.

> posed regardless of this separation. Yet to what extent as the same? I can, I must ask in this manner; for in another respect it [the "I"] is opposed to itself.[25]

Oedipus' moment of determinability, then, reminds us of Actaeon looking in the water and seeing a deer. *Who he is* becomes something that appears to Oedipus phenomenally in that moment of passage when he must ask, *who am I*? And in that same stroke, he feels the temporal tensions between immediately successive states of affairs that are contracted into one instant by simultaneous combining and dividing forces.

Now, the caesura is a cut for Hölderlin not just in the sense of dividing the play's rhythmic pace; it also cuts into its 'melody' or its 'rhyme scheme' so to speak. We would normally expect Oedipus as a tragic character to die from a vengeful murder, which would thereby initiate a new tragic cycle, like the deviant but cyclical paths of the planets. Instead, Oedipus' life is spared, and in *Oedipus Coloneus*, we see that he is left to aimlessly wander blind, his death indefinitely delayed. He is like a planet whose orbit was bent out of its circular course and will never return to where it started. Prior to and following that caesura, his life does not rhyme, and it is as if he is suspended between the *before* and *after* of this transition, that is, within the pure and empty form of time.[26] Thus, we can see the important difference between Merleau-Ponty's Swann and Deleuze's Oedipus illustrations. Swann seemingly came to differ from himself over an extended period, when in fact he remained essentially the same the whole time. Merleau-Ponty's model of temporal selfhood recognizes that we are continually differing from ourselves, yet we are always foretold of our changes, because our forthcoming states are implied in our prior ones. Oedipus, however, had no protentional awareness of himself being his father's murderer and his mother's lover, because he had sufficient cause to think otherwise. Up until the prophecy it was not even implicitly on the horizon of his awareness, hence the sudden shock of his realization.

Deleuze further elaborates this point with Shakespeare's story of Hamlet, whose life as well does not 'rhyme.' Hamlet is a thoughtful and contemplative prince; he is certainly not one to spontaneously seize action and murder someone. But, Hamlet's father, the former king, appears to him as a ghost and tells Hamlet to avenge his death. The former king's brother, Claudius, killed him and married his wife the queen, which prevented the crown from being passed on to Hamlet. After this conversation with the ghost, Hamlet says "the time is out of

25 Hölderlin, *Werke*, 216–217/37–38.

26 *Ibid.*, 202/108; Jean Beaufret, "Hölderlin et Sophocle," preface to *Hölderlin: Remarques sur Oedipe. Remarques sur Antigone*, by Friedrich Hölderlin, trans. François Fédier (Paris: Union générale d'éditions, 1965), 18.

joint." This could be because he now sees that the lawful cycles of kings living their full lives and bequeathing the throne to their first born son has been disrupted by the brother's murderous act. Yet, time is out of joint for Hamlet also because he now must make a passage between incompatible appearings of himself, at one moment a gentle thinker, at another a violent assassin. So before and after the caesura (his sea voyage), the time in Hamlet's life does not rhyme and instead flies off its hinge.[27]

Shortly committing himself to murdering his father, Hamlet ponders:

> To be, or not to be–that is the question:
> Whether 'tis nobler in the mind to suffer
> The slings and arrows of outrageous fortune
> Or to take arms against a sea of troubles
> And by opposing end them.

Hamlet can either keep himself living throughout a life of misfortune or take his life and spare himself from that suffering. Now on the basis of Deleuze's illustrations, I will suggest another example in hopes that readers may find for themselves similar experiences for their own phenomenal analysis. So if you wish, consider how your first time jumping into deep water might have filled you with intense trepidation. You were not yet a person who was able to swim, therefore the person you are when first making the leap is risking self-inflicted death. Yet you will survive, if when hitting the water you cease being a frightened non-swimmer and instead become a capable swimmer. In that moment of the leap, you know you are not a swimmer, because you have not yet made contact with the water and begun swimming; however, by committing to the jump, you determine and affirm yourself as a swimmer at the same time, for otherwise you would not have a sane motivation for taking the lethal risk. In that passage between states, you embrace your determinability, the pure order of time that makes you always before and after yourself. The 'to be or not to be' of your indecisiveness turns into a 'to be *and* not to be' at the moment of incompatible self-determination: to be a swimmer *and* not to be a swimmer, to be Hamlet *and* not to be Hamlet, to be you *and* not to be you. Deleuze's literary examples, then, invite us to consider experiences when we temporally differentiate from our-

27 Henry Somers-Hall and Keith Faulkner each explain Hamlet's character change with reference to one of Deleuze's influences, Harold Rosenberg, who theorized on the sudden shifts in a theatrical character's identity. Somers-Hall's work is a notable source for the fundamental philosophical concepts involved in the third synthesis, and Faulkner's book is unique in its particular attention to related psychoanalytic issues. Keith Faulkner, *Deleuze and the Three Syntheses of Time* (New York: Peter Lang, 2006); Henry Somers-Hall, "Time Out of Joint: Hamlet and the Pure Form of Time," *Deleuze Studies* 5 (2011).

selves by affirming that who we have been so far is an aspect of our selfhood, while at the same time affirming that the contrary self we are now becoming is just as much an expression of who we are. The phenomenological question we ask is, in these experiences when all of a sudden we change drastically, are we aware of the tension within us as being a temporal tension? For, it is time that allows us to differentiate from ourselves while in the same stroke gluing our temporal variations to one another. And so in these instances, we wonder, does it become apparent to us how time is an essential part of who we are?

6. Conclusion

We see then how such moments of self-appearing are fundamentally unlike Merleau-Ponty's model of the organic integration of phenomenal horizons. For, in such self-splitting moments like these when we affirm our determinability, the self who we just were, lingering still in our retentional horizon, does not assimilate with the self we are now becoming, looming ahead in our protentional awareness. The *now* no longer integrates with the *next*, and there is a shock in the flow of time. So for Deleuze, the form of time – the ordinal relation of the *before* with the *after* – comes about because we regard ourselves as being different from ourselves while still these alien self-variations remain intimately bound to one another in that moment of passage.[28] And in these moments, *who we are* and as well the time of our lives both become phenomena in our explicit awareness. For example, when we look in the mirror and abruptly notice a sign of age, we might at first not recognize ourselves, and we also see that years have passed right before our eyes. But, this passed time does not appear phenomenally as a continuous thrust. It is a shocking difference, a cut that both divides while forcing together the prior with the forthcoming. Yet, how can we even begin to conceptualize the way that moments pass continually one-to-the-next, if every instant is a cut in the flow?

---

28 My account here of Deleuze's third time synthesis regards it in somewhat static logical terms, namely, as the temporal coincidence of incompatible states of affairs contracted together into a single instant. So, I do not regard the synthesis primarily as a process but rather as a logical situation. James Williams, however, offers a compelling process philosophical interpretation of Deleuze's time syntheses, in Williams, *Gilles Deleuze's Philosophy of Time: A Critical Introduction and Guide* (Edinburgh: Edinburgh University, 2011). I elaborate more on the non-process nature of my interpretation in Shores, "In the Still of the Moment: Deleuze's Phenomena of Motionless Time," *Deleuze Studies* (2014), forthcoming.

Following Daniela Voss, we might conceive the temporal caesurae in terms of Dedekind cuts,[29] and in fact Deleuze does give a mathematical explanation of Dedekind cuts in one of his course lectures where he discusses the points of caesurae in Sergei Eisenstein's montage sequences.[30] Dedekind set out to give a rigorous definition of numerical continuity. He noted that not all the points along a straight number line correspond to a rational number, because there are also among them an infinity of irrational quantities, like *pi* and the golden ratio for instance. If we tried to render them into numerical digits, their decimals would never terminate, so in a sense they are found within infinitely small gaps posed between immediately successive rational numbers. Dedekind then defines the continuity of the real numbers as being a series of cuts that lie upon either a rational number or upon an irrational number that is situated between two successive rationals. Accordingly, we might then think of time as a series of cuts, with some of those cuts being 'irrational' gaps in time. Like moments of caesura, they in one sense are found along the continuity of temporal moments, while in another sense they fall outside the sequence. Thus, all the moments of our lives can pass by in apparent continuity except for instances when we find ourselves in a shocking gap interposed between *before* and *after*, where *our life as it was* and *our life to come* do not 'rhyme'. Our selfhood in this sense could be seen as the totality of all the cuts, of all the differences within ourself and not as the common denominator among all our variations or as the chain of local similarities bridging one moment to the next. And so we ask: which phenomenal experiences of time are more memorable? Is it when we sense time as a continuous thrust of a self that is organically integrated from moment to moment, or when we are shocked by who we are becoming in a moment that seems to stand outside all the rest? The self and time, it would seem, are more phenomenally given when they carry with them a differential tension. In fact, often times the more familiar something becomes, the less we notice it, like how after many repetitions we can commute to work without explicitly attending to very much along the way. And when seeing ourselves in the mirror each morning, so long as we do not notice anything different about ourselves, we can fix our hair or clothes

29 Daniela Voss, "Richard Dedekind's Impact on Deleuze's Static Synthesis of Time" (paper presented at the "Deleuze, Philosophy, Transdisciplinarity" conference at Goldsmiths College, University of London, 12 Feb 2012); "Deleuze's Third Synthesis of Time," *Deleuze Studies* 7 (2013); *Conditions of Thought: Deleuze and Transcendental Ideas* (Edinburgh: Edinburgh University, 2013). These works by Voss as well discuss the many various other topics surrounding Deleuze's concept of the third synthesis of time, and they are unique in their careful and elaborative attention to the related mathematical notions.

30 Deleuze, *Cours 27/11/1984*, www2.univ-paris8.fr.

without *who we are* coming to our attention. Thus, Merleau-Ponty's model is perhaps more suited for explaining how our temporal selfhoods *dephenomenalize*, that is, become less apparent in our focal awareness. So while Merleau-Ponty, like Deleuze, sees time and self as involving self-affective self-differentiation, his organic conception of integrated temporality does not explain the phenomenality of our self-givenness in those startling and shocking moments when we appear to ourselves as different from ourselves. For this reason, I recommend as a result of these analyses that phenomenological studies of the self and time also take into account their differential composition.

References

Beaufret, Jean. "Hölderlin et Sophocle." Preface to *Hölderlin: Remarques sur Oedipe. Remarques sur Antigone*, by Friedrich Hölderlin. Translated by François Fédier. Paris: Union générale d'éditions, 1965.

Deleuze, Gilles. *Cinéma 2: L'image-temps*. Paris: Les éditions de minuit, 1985.

— *Cinema 2: The Time Image*. Translated by Hugh Tomlinson and Roberta Galeta. Minneapolis: University of Minnesota/Athlone, 1989.

— *Cours 14/03/1978*. Translated by Melissa McMahon. www.webdeleuze.com.

— *Cours 21/03/1978*. Translated by Melissa McMahon. www.webdeleuze.com.

— *Cours 20/01/1981*. www2.univ-paris8.fr.

— *Cours 12/04/1983*. www.webdeleuze.com/www2.univ-paris8.fr.

— *Cours 13/12/1983*. www.webdeleuze.com/www2.univ-paris8.fr.

— *Cours 17/04/1984*. www2.univ-paris8.fr.

— *Critique et Clinique*. Paris: Les éditions de minuit, 1993.

— *Essays Critical and Clinical*. Translated by Daniel W. Smith and Michael A. Greco. Minneapolis: University of Minnesota Press, 1997.

— *Différence et répétition*. Paris: Presses universitaires de France, 1968.

— *Difference and Repetition*. Translated by Paul Patton. New York: Athlone, 1994.

— *Kant's Critical Philosophy: The Doctrine of the Faculties*. Translated by Hugh Tomlinson and Barbara Habberjam. London: Athlone, 1995.

Deleuze, Gilles and Félix Guattari. *Qu'est-ce que la philosophie?* Paris: Les éditions de minuit, 1991.

— and Félix Guattari. *What Is Philosophy?* Translated by Hugh Tomlinson and Graham Burchell. New York: Columbia University Press, 1994.

Faulkner, Keith. *Deleuze and the Three Syntheses of Time*. New York: Peter Lang, 2006.

Hölderlin, Friedrich. *Sämtliche Werke, vol. 5*. Stuttgart: W. Kohlhammer, 1952.

— *Essays and Letters on Theory*, edited and translated by Thomas Pfau. Albany: SUNY, 1988.

Husserl, Edmund. *Zur Phänomenologie des inneren Zeitbewusstseins*, *Husserliana X*, edited by Herman Leo Van Breda. Den Haag: Martinus Nijhoff, 1966.

— *On the Phenomenology of the Consciousness of Internal Time* (*1893 – 1917*), *Edmund Husserl: Collected Works, vol. IV*, edited by Rudolf Bernet. Translated by John Barnett Brough. Dordrecht: Kluwer, 1991.

Kant, Immanuel. *Kritik der reinen Vernunft, Erster Teil, Werke Vol. 3*, edited by Wilhelm Weischedel. Darmstadt: Wissenschaftliche Buchgesellschaft, 1968.

— *Critique of Pure Reason.* Translated by Paul Guyer and Allen W. Wood. Cambridge: Cambridge University, 1998.

Leibniz, Gottfried. "Theoriae motus abstracti Definitiones." In *Die Philosophische Schriften, vol. 4*, edited by Carl Immanuel Gerhardt, 227-228. Hildesheim/New York: Geog Oms, 1978.

— "The Theory of Abstract Motion: Fundamental Principles." 1956. In *Philosophical Papers and Letters*, edited and translated by Leroy E. Loemker, 139–141. Dordrecht: Kluwer, 1989.

Merleau-Ponty, Maurice. *Phénoménologie de la perception.* 1945. Paris: Gallimard, 2006.

— *Phenomenology of Perception.* 1962. Translated by Colin Smith. London/New York: Routledge, 2002.

— *Le visible et l'invisible.* 1964. In *Œuvres.* Edited by Claude Lefort. Paris: Gallimard, 2010.

— *The Visible and the Invisible*, edited by Claude Lefort. Translated by Alphonso Lingis. Evanston: Northwestern University, 1968.

— "Le Cinéma et la nouvelle psychologie." *Les temps moderne* 26 (1947): 930–947.

— "Film and the New Psychology." In *Sense and Non-Sense*, translated by Hubert L. Dreyfus and Patricia Allen Dreyfus, 48–59. Evanston: Northwestern University, 1964.

Ovid. "The Transformation of Actaeon into a Stag." In *Metamorphoses.* Translated by Joseph Addison. Ware: Wordsworth, 1998.

Rimbaud, Arthur. *Complete Works, Selected Letters: A Bilingual Edition.* Translated by Wallace Fowlie. Chicago: University of Chicago, 1966.

Shores, Corry. "Body and World in Merleau-Ponty and Deleuze." *Studia Phaenomenologica* 12 (2012): 181–209.

— "In the Still of the Moment: Deleuze's Phenomena of Motionless Time." *Deleuze Studies* (2014) [forthcoming].

Somers-Hall, Henry. "Time Out of Joint: Hamlet and the Pure Form of Time." *Deleuze Studies* 5 (2011): 56–76.

Voss, Daniela. "Richard Dedekind's Impact on Deleuze's Static Synthesis of Time." Paper presented at the "Deleuze, Philosophy, Transdisciplinarity" conference at Goldsmiths College, University of London, 12 Feb 2012.

— "Deleuze's Third Synthesis of Time." *Deleuze Studies* 7 (2013): 194–216.

— *Conditions of Thought: Deleuze and Transcendental Ideas.* Edinburgh: Edinburgh University, 2013.

Williams, James. *Gilles Deleuze's Philosophy of Time: A Critical Introduction and Guide.* Edinburgh: Edinburgh University, 2011.

Corry Shores
Department of Communication and Design
Bilkent University
Bilkent, Ankara 06800
Turkey
corryshores@gmail.com

Cruelty and Melancholy: The Stones of Roger Caillois

Josef Hrdlička

Abstract

The essay compares the conceptions of reverie of Gaston Bachelard and Roger Caillois in his texts about stones. Caillois in contrast with Bachelard doesn't emphasize the separation of rationality and reverie, rather he considers the latter as a continuation of ratiocination. One of the central analogies between both authors is the influence of the object upon the dreaming self. Bachelard speaks about an "infusion of being" by the image/object. It is possible to describe this determination of the self by the object along a scale of affectional states which are not only attributes of the one Self but create multiple Selves. There are two important affectional stances in Caillois's reveries: meditation transcending human time, in which consciousness escapes from the body and is cast back into it afterwards; and the body as it instrumentally acts upon the stones, removing them from their original hiddenness—cutting, grinding, polishing, which are all attempts to penetrate and intrude upon the stone. The shapes and designs of the stones seem in this respect to be structures to which man always contributes even though he is not their creator. On the level of inevitable interaction between man and stones, Caillois arrives at a "diagonal" analogy, accompanied by the harmonization of "elevated melancholy" (Ficino, Starobinski).

Keywords: Stones: Description and Aesthetics, Stones and Man, Stones and Myth, Melancholy, Roger Caillois, Gaston Bachelard

I

In December 1934, not long after leaving Czechoslovakia, Roger Caillois broke with the Surrealists. Several participants gave different accounts on different occasions of this event, and while it may appear somewhat humorous to us today, there is a serious point:

> Someone brought a few of those jumping beans from Mexico that are sold in souvenir shops. Breton wanted us to dream, allowing the object to move us into trance. I wanted to open one of the beans and find out if there was an insect or larva inside, as indeed there was. The question suddenly became decisive for me. It was the proverbial last drop. The next day I left the group and without delay I sent Breton a letter that I published soon after.[1]

1 "Quelqu'un avait apporté du Mexique plusieurs de ces haricots sauteurs qu'on y vend dans les boutiques de souvenir. Breton voulait qu'on rêvât, qu'on s'extasiât sur le prodige. Je préconisais qu'on ouvrît plutôt une des graines pour vérifier si un insecte ou une larve n'y serait pas contenu, ce qui (par parenthèse) était le cas. La question pour moi

Caillois's precipitate departure, as well as the letter, was the typical Surrealist note of the times. Although his gesture was not a complete break—Breton and Caillois remained in communication and continued to be respectful of one another—there was a fundamental disagreement about Surrealist poetics, which the scientifically inclined Caillois found too irrational, despite his attraction to the movement. Such ambivalence is also evident in his attitude to poetry. From a young age he wrote poems and edited several anthologies, but his writings are sceptical about modern poetry, as represented precisely by Surrealism. A work entitled "The Situation of Poetry" (*Situation de la poésie*) begins in this way: "To speak personally, I always felt more inclined to do battle with poetry than to give in to it. Oh, those were the days when, almost without exception, only lyric poems were published—and lyricism in no way pleases me. My view was that the artist should rather hide behind his work."[2] Caillois wrote poems very seldom, but his writings on the subject of stones are fittingly published and read as poetry. Moreover, he does exactly what Breton demanded—he dreams about objects.[3]

To return to the beans: as the subject of dreaming they are at once worthwhile and full of contradictions. If Caillois were to cross-section the object he would remove its momentary aura of the miraculous, which it retains as long as we don't know there is a jumping bean moth (*cydia deshaisiana*) inside it. However, once we know this, our dreaming, which is dependent on our ignorance, loses its necessary aleatory force. This Surrealist confrontation reminds us somewhat of Bachelard's typology of "perspectives of the hidden" and reveals one of the problematic points of his poetics. In *Earth and Reveries of Repose* (*La terre et les rêveries du repos*) Bachelard distinguishes four different views of the "interior" of matter, or the object.[4] Of relevance for us are the second and

était soudain devenue décisive. C'était la goutte d'eau classique qui fait déborder la vase. Dès le lendemain, je rompis avec le groupe et publiai peu après la lettre que j'avais envoyée sur-le-champ à André Breton." Roger Caillois, *Œuvres* (Paris: Gallimard, 2008), 214; henceforth only *Œuvres*.

2 "Si c'est ici l'occasion d'une confidence, je me suis toujours senti plus disposé à combattre la poésie qu'à m'y abandonner. Voici beau temps que les poèmes qui paraissent sont presque exclusivement lyriques et le lyrisme ne me plaît guère. J'ai l'idée que l'artiste doit se dissimuler davantage derrière son œuvre." *Œuvres*, 345.

3 Roger Caillois's main works about stones, from 1960 on, are reviewed by Juan Rigoli in "Caillois, le lapidaire intime," *Mémoire et oubli dans le lyrisme européen. Hommage à John E. Jackson*, ed. Dagmar Wieser and Patrick Labarthe (Paris: Honoré Champion, 2008), 426. In this essay, for the most part I draw upon the collection *Pierres, suivi d'autres textes* (1966) and the later essay *Récurrences dérobées. Le Champ des signes* (1978) as they are given in the publication of Callois's collected writings: *Œuvres*.

4 Gaston Bachelard, *La terre et les rêveries du repos* (Paris: Librairie José Corti, 1997), 9.

third: the dialectic perspective dreams in contradiction to the surface of phenomena (for instance, the black colour of milk, the large dimensions of a small object);[5] the perspective of amazement finds the miraculous interior of matter in a manner reminiscent of the work of human art (here Bachelard introduces examples from mineralogy).[6] For our purposes the relevant distinction is that from the outset the dialectical perspective penetrates matter with the help of dreaming and thus can be designated as poetic, while the perspective of amazement dreams upon the known "interior" of matter, for instance, crystals hidden in a geode; this second type of dreaming is thus in essence proto-scientific.

Bachelard in fact never strictly distinguished these two types of dreaming. He does not say which is dreaming from its outset and in its nature, and that which is mixed with rationality and unreflectingly creates with it a combination of these (although it is just this type of dreaming which led him to the strict division of imagistic and rational thinking). Bachelard provides examples from both areas in his last books. Roger Caillois is, on the other hand, rather the critical successor of Bachelardian proto-science instructed by modern thought. He was able to critically distinguish these three possibilities (that is, pure rationality, pure dreaming and the combination of both); nevertheless he understood as originary the connection of rationality and dreaming that Bachelard attempted to divide: "The scientific spirit appears to be a delicate composition of exact observations and sometimes daring analogies, or, in other words, critical caution and imagination."[7]

It is however important to note that we are dealing here with natural objects. A bean with a moth inside is not a secret cabinet, which is a fitting dream object.[8] It is a natural object and the phenomenon can be explained; at least for Caillois, the wonderful as a category begins when what can be revealed is revealed, and is a continuation of rational thought rather than a divagation from it. Dreaming which penetrates the depths of natural phenomena is different from dreaming about a created object and resembles more the second of Bachelard's

5 *Ibid.*, 12–28.

6 *Ibid.*, 28–33.

7 "L'esprit scientifique semble une composition délicate d'observations exactes et d'analogies parfois aventurées, ou, si l'on veut, de vigilance critique et d'imagination." *Œuvres*, 1162; cf. also 1147: "My extravagant reflection again attempts to present a physical phenomenon and the tendency to imagination as equivalents in their respective milieu." ("Ma réflexion extravagante s'emploie une fois de plus à présenter comme équivalents dans leur milieu respectif un phénomène physique et une pente de l'imagination.")

8 Cf. chapter entitled "Drawers, Chests and Wardrobes" in Bachelard's *The Poetics of Space*, trans. Maria Jolas (Boston: Beacon Press, 1994), 74–89.

abovementioned perspectives. Analogically Caillois, from the beginning of his association with the Surrealists, considered psychological automatism as if it were a natural phenomenon and not a subjective activity. He appreciated the method as a foundation, but he considered it necessary to analyse and assess its products:

> I naively thought that Surrealism was not a literary movement like all the others but rather proclaimed the end of all literature. I had thought that it had resolved to substitute literature with the rigorous study of the imagination, notably through automatic writing, defined, according to the famous formulation in Breton's Manifesto, to reveal the true functioning of thought beyond all moral, intellectual and aesthetic control.[9]

We can summarize the position thus: modern poetry is too subjective, too given over to the search for original images;[10] if there is something important for Caillois in this poetry, it is not subjective expression but rather an objective structure of thought that can be studied to advantage. Poetry thus shows itself to be not the subjective product of a single person, but a phenomenon, if not natural, then at least arising from a collective psychological basis. For Caillois, an individual's creative output is too arbitrary, and he favours a reflective subjectivity close to that of the scientific mind rather than poetic spontaneity.

II

Bachelard remarked that dreaming defines the *cogito* of the dreamer,[11] the dream object forming a world and also marking the dreamer: "Thanks to a fruit, the whole being of the dreamer becomes round."[12] The same is true of Caillois's stones. He expresses the complex transactions between the dreamer and the dreamed thus:

---

9 "J'avais été naïvement persuadé que le surréalisme, loin d'être un mouvement littéraire de la même nature que les autres, proclamait au contraire la fin de toute littérature. Je pensais qu'il se donnait pour tâche de la remplacer par l'étude rigoureuse de l'imagination, au moyen notamment de l'écriture automatique, destinée, selon la formule célèbre du *Manifeste* de Breton, à révéler le fonctionnement réel de la pensée en dehors de tout contrôle moral, intellectuel ou esthétique." *Œuvres*, 214.

10 Cf. Caillois, "Aventure de la poésie moderne," in *Approches de la poésie* (Paris: Gallimard, 1978), 51–64.

11 Gaston Bachelard, *Poétique de la rêverie* (Paris: Presses Universitaires de France, 1989), 132 ; *The Poetics of Reverie: Childhood, Language and the Cosmos*, trans. Daniel Russell (Boston: Beacon Press, 1969), 154.

12 Bachelard, *The Poetics of Reverie*, 154. "Grâce à un fruit, c'est tout l'être du rêveur qui s'arrondit." *Poétique de la rêverie*, 133.

> When I regard stones attentively, now and again I not unnaively try to divine their secrets. I let myself imagine how such enigmatic marvels were formed, created according to laws that they very often seem to break, as though they came out of a tumult, or rather celebration, which their mode of existence henceforth eliminates. I try to seize them in my thoughts at the burning instant of their genesis. At such moments, I experience a strange sort of excitement. I feel that I share in small part something of the nature of the stones. At the same time, I bring the nature of the stone close to my own nature thanks to unsuspected attributes that strike me as applicable to them during the course of my thought—now exact, now loose—in which the warp of dream and weft of knowledge are composed. Here delicate constructions, perhaps integral to thought, ceaselessly rise and fall. Metaphor here denies (or disturbs) syllogism; vision nourishes (or distracts) rigour. Between the fixity of stone and the effervescence of thought a kind of current is established, in which for a short but memorable moment, I find wisdom and reassurance.[13]

Bachelard distinguishes the poetic *cogito* from Descartes's (avoiding the heart of Cartesian thought) when he says that he is not concerned with the *ego cogito* as something stable and certain. Instead he turns to the object (*cogitatum*) and its influence on the human subject: "Immediately, reverie has a simple object, the friend and companion of the dreamer."[14] The structure is reminiscent of the phenomenological intentional schema of *noesis-noema*, but here too the argument moves in a different direction. Passing over the many differences, the key point is that dreaming is not founded on the identity of the *I*, but rather it is as though the object of dreaming always creates a new and qualitatively distinct *ego cogito*. Certainly, it is still I myself who is dreaming, but the temporarily dreaming *cogito* is so forceful that there is not only a change of intentionality, but also of

---

13 "Lorsque je regarde attentivement les pierres, je m'applique parfois, non sans naïveté, à en deviner les secrets. Je me laisse glisser à concevoir comment se formèrent tant d'énigmatiques merveilles, nées de lois que très souvent elles paraissent violer, comme si elles étaient issues d'un tumulte et, pour tout dire, d'une fête que bannit désormais leur mode d'existence. Je m'efforce de les saisir en pensée à l'ardent instant de leur genèse. Il me vient alors une sorte d'excitation très particulière. Je me sens devenir un peu de la nature des pierres. En même temps, je les rapproche de la mienne grâce aux propriétés insoupçonnées qu'il m'arrive de leur attribuer au cours de spéculations tour à tour précises et lâches, où se composent la trame du songe et la chaîne du savoir. Là s'échafaudent et s'écroulent sans cesse de fragiles édifices, peut-être nécessaires. La métaphore y épaule (ou y corrompt) le syllogisme; la vision nourrit la rigueur (ou la fourvoie). Entre la fixité de la pierre et l'effervescence mentale, s'établit une sorte de courant où je trouve pour un moment, mémorable il est vrai, sagesse et réconfort." *Œuvres*, 1078.

14 Bachelard, *The Poetics of Reverie*, 22. "Tout de suite, la rêverie a un objet, un simple objet, ami et compagnon du rêveur." *Poétique de la rêverie*, 20.

existential definition: the image "infuses us with being,"[15] newly defining who I am. In his texts about stones, Caillois opens up a whole spectrum of such subject positions, given in relation to the stone. How can we understand such existential definition? Is this not merely a dreamed, fictional subject that disappears as soon as the dreamer moves the stone away or the reader closes the book?

The infusion of being that Bachelard mentions can be characterized in various ways. With respect to the subject the key element is emotionality—emotions as states that, unlike sensory input, are not subordinate to the self, but transform it, attune it, move it and disturb it. Which in Caillois, preoccupied with the object more than Bachelard, is more apparent: "At such moments, I experience a strange sort of excitement." The infusion of being is a mood, an emotion or passion which engulfs me; at that moment I am truly someone else; and this then abates, or more exactly, another disposition comes on, which may be more neutral or less emphatic. These dreamed or literary emotions are in the first degree just as real as others, seemingly differing from the moods of everyday life only by virtue of their unpragmatic nature. They equally move me or carry me away; but they do not provoke me to immediate physical reaction. And they equally engage in the "life" of my subjectivity, leaving traces after they pass that may be negligible or more marked: "I would however be left with the memory, a splinter in the skin, something miniscule that persists. For it is no small thing, such a silence in the uproar, in the unremitting and unchanging discourse that engulfs our days."[16]

III

Caillois's meditation on stones repeats and varies different motifs and modes of description and reflection, and these develop over the course of his entire oeuvre. Nevertheless, I hold that his texts about stones do not lead to a single, unambiguous position or conclusion—rather repetition itself is characteristic.[17] If we start from stones as objects of dreaming it is possible to speak of a space of the imaginary that dreaming constitutes around them; and one of the ways we can read this space is through the range and character of emotions or effects it accommodates. Within this spectrum I will attempt to describe the genesis of

15 Bachelard, *The Poetics of Reverie*, 153. "Elle [image] nous infuse de l'être." *Poétique de la rêverie*, 132.

16 "Il m'en resterait toutefois un souvenir, une écharde, quelque chose d'infime qui demeure. Car ce n'est pas rien que pareille bonace dans le vacarme, dans le discours sans trêve ni virgule où s'engloutissent nos jours." *Œuvres*, 1079.

17 Rigoli discusses the variability of these writings, "Caillois, le lapidaire intime", 426–427.

melancholy from the initial feeling of fascination, while also drawing attention to the circular structure in the background.

The first type of emotion that the stone evokes is the feeling by which it attracts attention, by which it first draws our consciousness, and instigates dreaming. The stone draws us by the surprising way it resembles something else. Caillois became interested in stones:

> [...] accidentally and through the comparison of different realms which attracted me strongly. I discovered in labradorite an optical colour (by which I mean neither chemical nor pigmental), an intense electric blue, which according to the angle of the light changed radically into a sandy grey. In my childhood village, I was amazed by the same blue that was capable of flipping over on the wings of a very common butterfly at that time, a Purple Emperor."[18]

Resemblance and analogy are then also the main ways of describing the stones. Seeing analogies becomes a passion: "I give myself over to the demon of analogy."[19] Here the unavoidable human tendency to seek something known in indefinite shapes manifests itself ("that imperious need to treat every accidental configuration as a rebus and to people it with the familiar images of clouds, the bark of trees and flaking walls"[20]).

At the same time Caillois is aware that analogy has two aspects: analogy that imagination introduces into the object, but also resemblance that cannot be dismissed as the mere work of dreaming. Both of these constitute the question of stones. Caillois explains them through the existence of laws common to both various natural domains and human culture. One of the examples is geometrical structures, for instance of crystals or patterns in stones which resemble geometry discovered by people.[21] These two types of analogy or resemblance cannot be clearly distinguished from one another. Caillois finds resemblances within the whole spectrum of nature, from plants to fauna and to people, as well as analogies with human culture, in the stones' shapes and above all in the drawings on their surfaces. In this respect stones represent a micro-world and dreaming about them is at once the discovery of the world and the revelation of deeper relations

18 "[A]ccidentellement et dans la perspective des rapprochements interrègnes qui m'attiraient avec une très haute priorité. J'avais retrouvé dans la labradorite la couleur optique (c'est-à-dire non chimique, non pigmentaire), un bleu intense, électrique, qui d'un coup bascule dans le gris poussière suivant l'angle d'incidence de la lumière. Dans le hameau de mon enfance, j'ai été émerveillé par le même bleu capable de chavirer sur les ailes d'un papillon alors assez commun, le Mars changeant." *Œuvres*, 147.

19 "Je me livre au démon de l'analogie." *Œuvres*, 1080.

20 "[C]et impérieux bésoin de traiter en rébus toute configuration accidentelle et de peupler d'images familières les nuées, les écorces et les murs craquelés." *Œuvres*, 1050.

21 Cf. the consideration of hyperbolas in *Le champ des signes*, in *Œuvres*, 1132.

than those with which science normally works. At some points resemblance is aleatory and does not repeat itself;[22] at others it shows itself to be part of the structure of the stone itself, exceeding mere figuration and leading to ratiocination about the laws of nature:

> Certainly, these salts are a mirage that perfectly mirror the vegetal even though they are spared both life and its extinction. However, I cannot but be convinced that these false ferns, mere resemblances of plants and pertaining to an incompatible realm, in their way alert the spirit to the existence of vaster laws that also govern the inert and inorganic.[23]

These laws, connected with Caillois's idea of diagonal sciences, cannot be verified and necessarily retain their secret character. The observer of stones, of nature and even of human society may note their manifestations, finding analogies between distant elements of various areas, but he cannot test them through experiment, as he can the classic laws of nature. And even if that possibility existed, it would necessarily elude rational evidence of the observer on the phenomenological level. In this respect the physical person is the final court of appeal, his evidence, intuitions and attunement.[24] In order to reveal these laws, imagination is necessary, and critical reflection supplements this by preventing immediate enthusiastic acceptance, always returning the human to its physical existence.

On a fundamental level, if one notes such analogies and laws, one has a choice between two positions: the first position emphasizes the body, the physical; the second is removed from the body (is ecstatic). Caillois as observer en-

---

22 "Le château", "Un cuivre", "Autre cuivre" in Caillois, "Pierres," 1095–1096, 1048–1051.

23 "Mirage assurément que ces sels qui affichent si parfaite simulation du végétal, quand ils sont soustraits tout ensemble à la vie et à la corruption. Toutefois je ne parviens pas à me défendre de la conviction que ces fougères fausses qui n'ont de la plante que l'apparence et qui appartiennent à un règne incompatible, à leur manière avertissent l'esprit qu'il est de plus vastes lois qui gouvernent en même temps l'inerte et l'inorganique." *Œuvres*, 1048.

24 Caillois remarks on the resemblance between patterns in stones and some fictive narratives: "Increasingly, I am less able to distinguish the two sorts of images—those that are frozen in stone and those that issue in fictional vapours. I know that they are formed by other means, though for the same reasons. I am persuaded that they partake in a fraternal destiny whose common root I must vainly try to discover. I suspect that it is not possible for localized, evanescent, if not superfluous, beings to reach this." ("Je réussis de moins en moins à distinguer les deux sortes d'images, celles qui sont gelées dans la pierre et celles qui sont issues des vapeurs de la fiction. Je sais qu'elles sont formées par d'autres voies, toutefois pour les mêmes raisons. Je me persuade qu'elles partagent une destinée fraternelle, dont je m'efforce en vain de découvrir la racine commune. Je devine qu'il n'est pas à la portée d'un être local, momentané, sinon excédentaire, d'y parvenir.") *Œuvres*, 1159.

counters his physicality at the moment when his body manifests itself as a limitation, as his meditation on stones fades. Even if by chance he

> once regained this unity, and acquired this detachment, a moment later I would be—neither more or less than the liberated others (than the spellbound others) in the dark nights and châteaux of the soul—restituted to a common state, to my place and to my temporary body, which is neither permanent nor lapidary.[25]

The body manifests itself in a very different affective way when the observer desires to penetrate the secret of stones: here she cuts them, smoothes them, polishes them and submits them to chemical processes.[26] The drawing on the stone is revealed only after the cut is made and then polished—it is not thus a normal phenomenon, but an image that the human frames. Without this violence, stones would sleep on "lodged in their dark lodes".[27] When they are cut, sight, and more generally visibility, enters the relation. Frequently a doubt arises whether all the resemblances and laws are not introduced into stones by humans themselves. As Caillois says on several occasions: yes, it is not so unambiguous and this noting of resemblances has two fundamental poles. Although the violent action that penetrates to the interior of the stone makes visible its drawing, that

---

25 "[L]'unité une fois retrouvée et le détachement acquis, je me retrouverais la seconde d'après ni plus ni moins que les autres délivrés (que les autres fascinés) par les nuits obscures et les châteaux de l'âme, restitué à la condition commune, à ma place et à mon corps passager, qui lui n'est ni durable ni pierre." *Œuvres*, 1079.

26 The motif of the cut and polishing occurs frequently. Finishing, or treatment, is the central motif of "Recipe" (*Recette*; 1098), where the pattern of stones comes forth only thanks to a series of ingenious mechanical and chemical interventions. A pattern resembling that found in stones, this time in wood, is not only frequent but necessary "only when a particular cutting technique brings it forth" ("pourvu qu'une précise technique de sciage les fasse apparaître", *Œuvres*, 1137). The essay on water in stones is also remarkable (*Œuvres*, 1062–1064). Water locked in some stones provokes the desire to penetrate to it inside, but when the stone is thus disturbed, the water evaporates. It came about before mankind and is perfectly unreachable by humans, thus it is locked in stone. Nevertheless it is accessible by osmosis: "We also find this captive water only in less porous substances, such as quartz or chalcedony, which completely, or almost completely, prevent any osmosis or transpiration. Chalcedony is not however an entirely secure prison since the able artisans between Eifel and Hunsrück can infiltrate it with colour." ("Aussi ne trouve-t-on cette eau captive que dans les substances les moins poreuses, comme le quartz ou la calcédoine, qui interdisent ou peu s'en faut toute osmose, toute transpiration. Encore la calcédoine n'est-elle pas une prison tout à fait sûre, puisque des artisans habiles entre l'Eifel et le Hunsrück parviennent à y infiltrer une couleur" [1063]). Music is discussed as a remarkable means of harmonic violence in one of his late writings about stones *Le Champ des signes* (1978).

27 "... dans leur gîte et la nuit des filons." *Œuvres*, 1037.

drawing is in its own way objective. The human creates the field of the visible, revealing structure, but does not know in advance what will come forth. In this act commerce takes place between the two sides, and further meditations and dreaming develop this. In contrast some resemblances arise only by chance, but the mere fact that the stone holds my attention opens the pole of the visible in which I have a relation to it, and this minimal relation cannot be gainsaid. The stone's hiddenness has been disturbed.

The texts about stones here display a noticeable contradiction which Caillois was evidently aware of, although he shifted it to other levels. The dedication of *Stones* on the one hand clearly speaks of hidden stones untouched by humans,[28] but speech itself makes of them a human work: speech now is a manifestation of that revealing and visualizing violence (which is of course violence only in a certain respect). As soon as the stone enters the sphere of the visible (either drawn in by violence, or itself drawing the attention of the human), it is no longer possible to differentiate the human observer and the inhuman mineral. Certainly it is possible to distinguish degrees, there being a difference between a stone that is untouched and merely observed, and one that is processed, that, for instance, is drawn upon (completely extant markings). However, this does not alter the fact that the beginning of the relation between the human and the stone prefigures this distinction, and the already foreshadowed "processing". We cannot unambiguously say "who started", whether the stone provoked the person to notice it or whether the person dragged the stone from its lodging by violence. (Here there is no question of anthropomorphism: the stone of course is not active, but *is*, and a physical person stands in the world somewhere beside it.) Caillois here arrives at the paradox of the "world" itself: a human being is in the world with everything that pertains to it, the world and some objects (here stones) that precede the human being; but neither does this world, as a world, exist without the human. The most distant correspondences that Caillois finds between the tendency to dream and physical phenomena perhaps arise because of the contiguity of humans and stones in the world.[29]

IV

For Callois, along the spectrum of emotions and positions that correspond to stones, the final one which all considerations reach is melancholy, which is pri-

28 "I speak of stone that always lay outdoors or of stones which sleep lodged in their dark lodes." ("Je parle de pierres qui ont toujours couché dehors ou qui dorment dans leur gîte et la nuit des filons.") *Œuvres*, 1037.

29 Caillois, *Le champ des signes*, 1147, 1149.

marily spiritual. This arises from the recognition that the inception of stones is unknowable for humans, that our imagination can only be approximate in this matter. Through its permanence and impenetrability, the unmoving stone suddenly outstrips the human and turns the meditating individual back upon him- or herself, recalling the temporary nature of his or her existence and producing a state of spiritual paralysis typical for melancholy. The introduction to *Stones* reminds us that stones precede human consciousness and will endure after the latter is gone; they exist in a dimension of time that significantly surpasses human time:

> I speak of stones older than life, stones that outlive life on cooled planets, if indeed there was ever life there. I speak of stones that do not even have to wait for death and that have no other work but to let sands, downpours, storms, time and tide break over their surfaces.[30]

Meditation on stones now results in a fairly sharp division of physical and spiritual movement between the two poles. The observer gains something from the immobility of the stone, but between the turbulence of thoughts and the stones a kind of current is established.[31] In this meditation, Caillois immerses himself in the depths of the stones' imagination and the active agent is the dreamer's awareness, which also brings the stone to life, while the human body is just as lifeless as the material of the stone. Here we have a type of description that differs from analogy: Caillois frequently considers stones from the perspective of their genesis, understanding them as the result of the work of primitive forces and energies.[32] The stone itself is in this respect the trace of a state that preceded it, a petrified record of something that no longer exists. At the same time it registers forces and events which, with the passage of time, produce the human being. The stone as a trace, as a mere record of something that preceded it, as a frozen image of a past event, prompts the imagination to conceive of how such traces originated.

Caillois dreams of regression to the beginning much as the poet Roger Gilbert-Lecomte does,[33] but with the important difference that he decouples the physical human being from the activity: while dreaming over stones his rigid

---

30 "Je parle des pierres plus âgées que la vie et qui demeurent après elle sur les planètes refroidies, quand elle eut la fortune d'y éclore. Je parle des pierres qui n'ont même pas à attendre la mort et qui n'ont rien à faire que laisser glisser sur leur surface le sable, l'averse ou le ressac, la tempête, le temps." *Œuvres*, 1038.

31 *Œuvres*, 1078.

32 *Œuvres*, 1046.

33 Stéphane Massonet points to this connection, *Les Labyrinthes de l'imaginaire: dans l'œuvre de Roger Caillois* (Paris: L'Harmattan, 1998), 285.

(stone-like) body remains apart as dreaming shifts to the space of the imaginary.[34] While in Lecomte the dreamer changes as he moves backwards, his body also regresses; in Caillois this person remains dreaming here, but through dreaming it is as though consciousness (spirit) momentarily escapes its physical existence and the flow of time. Now we have emotion or attunement of a different degree: the feeling of abstraction, or ecstasy, provokes nostalgia for evanescent moments, but at the same time leaves the person with a lasting memory. Jean Starobinski characterizes the attunement of Caillois's later texts according to Ficino as a noble melancholy ("mélancolie généreuse") in which sadness is joined with the happiness of contemplation, lyric transport with melancholic return to the self.[35]

Genetic description of stones includes one of the explanations of analogies between the manifestations of various levels of the world or realms of nature, and it does this by following the activity of the forces or laws at work in these realms. One of the main features of Caillois's descriptions of stones is the motif of writing.[36] On a whole range of stones we find drawings resembling writing, which is not however writing, because the stones precede human beings. What is fascinating here is that the human, much later than these stones, but in the same fashion, used similar graphic designs and similar styles of drawing to make records. It is remarkable that they chose the same device from the repertoire of the finite world. In a text entitled "The Chinese Sign," the drawing on the stone reminds him of an unknown Chinese ideogram. Although the resemblance is deceptive (as it can't be writing), humans later repeat the shape that was in the stone when they use it as a sign. Caillois allows the mineral to speak thus:

> As for my form, it does not need to be decoded. It was never intended to express the least message. It never required more or less than a chain of chance to be one day discovered in the interior of a stone and revealed to sight. From its random resemblance to one of the writings invented by ingenious humans, it draws a strange splendour. Therefore it is necessary to admire it, and be astonished by the way that this form could precede such lettering. However it did not precede; neither did this lettering catch up with it or imitate it. Rather, it repeated my form. The patience of the world outstrips the flash of a thought, of a life, of an era. The miracle resides in

34 Elsewhere Caillois cites Charles de Bouëlle's tract *Liber de sapiente* (1510), where extreme grief, *accidie*, is connected with minerals (*Œuvres*, 1115). Worthy of remark here is the variability of meditation according to the object. In Bachelard we find the body of the dreamer similarly cordoned off from fire, or held up from the water's surface; however, the mood and atmosphere of the situation differs with respect to the character of the element.

35 Starobinski, "Saturne au ciel des pierres", 178.

36 For example, *Œuvres*, 1091, 1094, 1105–1107.

> an encounter that all conspires to render inconceivable, above all, the opposition of realms and processes. If I merit attention, it is because I am to be found on a kind of crossroads of an impenetrable network. It is not my place to judge this. That is a matter for the living who have nothing in common with me and who lend me a voice in play or by ruse. They pretend that they think and meditate upon me; but in reality they are thinking about themselves and they are the ones who are reflected in my image. Neither they nor I know the intention that led to such artifice. I suspect that they, harried and voluble, like to think they are kindred with me, who is a stone and a drawing on a stone, condemned to silence.[37]

Even these analogies can be characterized as slanted or diagonal because they manifest themselves in unsymmetrical levels and have another, impenetrable significance. They reveal the structure of the "finite world":[38] "Writings of stones: structures of the world."[39]

In the construction of the world, which is on an analogical level captured already in stones, the last component, "the topmost foam,"[40] is human imagination. Here we arrive at Caillois's melancholy: nothing in this construction of the world points to the human consciousness, or spirit, that hovers and floats over the stones. All the elements, including writing, human powers and faculties,

---

37 "Ma forme, quant à elle, n'exige pas d'être déchiffrée. Elle n'a jamais servi à exprimer le moindre message. Il ne lui a fallu rien de moins ni rien de plus qu'un chapelet de hasards pour être un jour découverte dans l'intérieur d'une pierre et proposée aux regards. Sur-le-champ, elle tira une étrange noblesse de sa ressemblance fortuite avec l'une des écritures inventées par l'ingéniosité des hommes. Dès lors, il en est pour admirer avec stupeur qu'elle ait pu les devancer. Elle ne les a pas devancés, ni eux ne l'ont rejointe ou imitée. Ils l'ont répétée plutôt. La patience du monde dépasse l'éclair d'une pensée, d'une vie, d'une ère. Le miracle est dans une rencontre que tout conspirait à rendre inconcevable, et d'abord l'opposition des règnes et des démarches. Si je mérite attention, c'est pour me trouver située à quelque carrefour de l'inextricable réseau. Il ne m'appartient pas d'en juger. C'est affaire au vivant qui n'a rien de commun avec moi et qui me prête une voix par jeu ou par ruse. Il feint de réfléchir, de méditer sur moi, mais c'est lui qu'il réfléchit, qu'il reflète sur mon apparence. J'ignore, il ignore sans doute aussi, dans quelle intention il eut recours à pareil artifice. Je soupçonne qu'il a souhaité affirmer, lui l'inquiet, le prolixe, sa complicité avec moi, qui suis pierre et dessin dans la pierre, condamné au silence." *Œuvres*, 1106–1107.

38 Thinking in the finite world almost necessarily leads to statements of resemblance: "Even though my examples seem aberrant or unconvincing, that does not mean that other, more convincing examples will not finally be found. In effect, research consists of a quasi inevitable conjecture in a finite world." ("Cependant, même si mes exemples semblent aberrants ou dissuasifs, il n'est pas exclu qu'on en découvre d'autres qui soient enfin plausibles. La recherche repose en effet sur une conjecture quasi inévitable dans un monde fini.") *Œuvres*, 1150.

39 "Écritures des pierres: structures du monde". *Œuvres*, 1094.

40 "… écume extrême". *Œuvres*, 1151.

even culture, are in stones "prefigured" in a certain evolutionary line: a device repeats itself later on, and thus retroactively finds its simulacrum in the stone. Even human imagination manifests certain tendencies, analogies for which can be found in the drawings on stones. The only thing that is not directly prefigured in stones is this last, reflective awareness, which is preoccupied with these very stones, and which finds itself on the borders of the world *qua* nature. Caillois's thought thus oscillates between a dualism and monism, and this apparent conflict is not resolved by the final texts, as Starobinski notes.[41]

V

One of these last texts is the study *The Field of Signs* (*Le champ des signes*). Here Caillois studies the "correspondences" apparent in the most extreme of possible ranges, going beyond apparent analogies, for instance between mineral and animal kingdoms, as well as the similarities between writing and drawings on stones, since he considers the resemblances between the pattern in the stones and some stories (*fable*). This is where it is possible to point to the proto-scientific imagination that Bachelard both criticizes and admires. Caillois is very much aware of certain pitfalls and keeps the examined phenomena at a proper remove, not shifting the phenomena to other levels, since he is thinking through the special connections between distant levels. Because of this, the text is also a meditation on the human, which is to be found somewhere in between. Caillois begins thus:

> I treat stones deferentially, but I take them for the insensible minerals they are and will remain. I take fables as fables, with the prudence, uncertainty and incredulity they require. However, more than once it struck me that stones could also be regarded as a kind of poetry, and that I might search for the permanence of stones, their immovable significance, in such fictions; in other words, try to reunite through some subtle feint the divided and contrasting parts of our indivisible universe.[42]

In this short introduction, he distantly alludes to the old myth of an ur-being whose body was divided in parts and from which the world emerged. Caillois

---

41 Starobinski, "Saturne au ciel des pierres", 184–185.

42 "Je traite les pierres avec déférence, mais en minéraux insensibles qu'elles sont et demeurent. Je tiens les fables pour fables, avec la prudence, l'incertitude et l'incrédulité qu'elles commandent. Plus d'une fois, cependant, il m'est arrivé de penser qu'il convenait aussi de regarder les pierres comme des sortes de poèmes et de chercher en revanche dans les fictions la pérennité des pierres, leur inébranlable signification, c'est-à-dire d 'essayer de réunir par quelque biais même ténu les parties disjointes et contrastées de notre indivisible univers." *Œuvres*, 1129.

does not mention the myth, but it is as though he attempts to find or renovate a similar shape of the world by a different route and on an unmythical level. From bone-stones to story-thoughts. As I remarked, his approach differs from Lecomte's and even Bachelard's, but they share the same goal.[43]

To summarize, in one quarry, Caillois found special quartzes with drawings resembling parallel bands which have the shape of a hyperbola. He finds similar drawings on other stones and elsewhere in the natural world, for instance in the growth rings in trees. He suggests that sound vibrations—musical effects on matter—originally brought about these drawings, and finds an analogy in Chladni Figures,[44] structures which occur in the fine dust of minerals subjected to sound vibrations.[45] The third part is a story:

> Thus far I have restricted myself only to stating that in the laboratory symmetrical compositions emerge at a remove or at least without direct contact by the action of appropriate vibrations. Thus it is not inconceivable that, transposing a similar phenomenon to an inaccessible tectonic, a daring mind will hazard a guess at the existence of miniscule, fitful waves, too slow for present technology to detect. The specific wavelengths, given power through repetition, could very well set up a ballet of dust, bringing it to a point of perfect equilibrium and leaving it to set in a definitive form. They are brought to a talc by topological powers of high speculation. They are moved as though by the notes of a magic flute, as are the children and the rats of Hamelin, where, one might say, a selective sorcery calls them, separates them, first gathering the rats together again and bringing them away, then the children and people, leading the first to the frozen river, and the others into a nearby mountain, which opens before them, and closes behind them again.[46]

---

43 Cf. Bachelard, "Man was the word of this macro-anthropos which is the monstrous body of the earth. In primitive cosmic reveries, the world is a human body, human eyes, human breath, a human voice." *The Poetics of Reverie*, 188. "L'homme était la parole de ce macro-anthropos qu'est le corps monstrueux de la terre. Dans les rêveries cosmiques primitives, le monde est corps humain, regard humain, souffle humain, voix humaine." *Poétique de la rêverie*, 161.

44 In 1787 Ernst Chladni (1756–1827) tried to vibrate level planes covered in the dust of various minerals; due to the movement, the dust formed symmetrical patterns dependent on the frequency and the shape of the planes. This is how Chladni produced tens of various images. Cf. *Œuvres*, 1143–1144.

45 *Œuvres*, 1144.

46 "Jusqu'à présent, je me borne donc à constater qu'en laboratoire, des compositions symétriques naissent à distance ou du moins sans contact direct par l'action de vibrations appropriées. Il n'est donc pas inconcevable que, transposant pareil phénomène dans l'inaccessible tectonique, un audacieux se hasarde à conjecturer l'existence de houles infimes et lancinantes, trop lentes pour être décelables dans l'état actuel de la technique. Des longueurs d'ondes spécifiques, dont la répétition fait le pouvoir, jadis ont pu aussi

In the second part of the essay Caillois is concerned with the story of the Pied Piper of Hamelin and compares it to two other stories in which a group of people, suffering from injustice and violence, leave the world for a sphere or realm somewhere apart, which is neither heaven nor hell. He considers their existence there as a constant dance, a harmonic movement incited by this magic. In their way, these beings come closer to nature, foregoing something of their human existence. Since through the inexorable logic of imagination, "[i]t is not possible to be freed [from human vicissitudes] and at the same time retain the privilege of breathing and being moved, to return the soul recently received. Thus these survivors are sharply segregated from their previous state, a state they will never regain."[47] Their perfect refuge is gained at the price of their human-ness, that which distinguishes them from nature. Caillois's analogy is thus especially dynamic, since a human being who returns to the harmony of nature in fact leaves the human story and merges into the natural pattern. These two strange myths partially supplement the analogy.

At this juncture Caillois is however at the very least cautious enough not to let himself be carried away by a simple resemblance. He declares that his examples are not inevitably always the most convincing—better ones could be found—but the main point is the necessity of bringing closer the phenomena of the world and imagination. Imagination hides in itself something of the secret of analogies. If we look closer at the beginning of *The Field of Signs*, once again there is a violent action, since the patterns in quartz mentioned above were revealed in a quarry, and thus brought to light by violence. But this act of force, in the course of the meditation takes on the more subtle character of a sound wave, of music that affects from afar, and things enter into the sphere of the visible not by violence, but as the result of a deeper movement whose nature is musical. Music connects the revelation of stones with its distant counterpart, the story of children locked in rock who are returning from the world of phenomena; from the sphere of the visible to the darkness and hiddenness of stone. Petrification

---

bien régler un ballet de poussières, le porter à son point de parfait équilibre et le figer dans une stabilité définitive. Elles ont amené un talc à des processes topologiques de haute spéculation. Elles ont agi comme les sons d'une flûte magique sur les rats et sur les enfants extraits successivement de Hameln, où, dit-on, une sorcellerie sélective appela, sépara, rallia, préleva enfin la troupe des rongeurs, puis les enfants des hommes et les entraîna les uns dans une rivière glacée, les autres dans les flancs d'une montagne voisine ouverte pour les accueillir et aussitôt refermés sur eux." *Œuvres*, 1144–1145.

47 "Il n'est pas possible de s'en [vicissitudes humaines] trouver affranchi sans renoncer du même coup au privilège de respirer et d'être ému, à la nécessité de rendre l'âme naguère reçue. Aussi une ségrégation décisive sépare-t-elle ces rescapés de leur ancien état, qu'ils ne retrouveront jamais." *Œuvres*, 1154.

here is an inversely symmetrical movement that closes the entire universe of Caillois's speculation. Between the lines he perhaps expresses the two-fold melancholy of the spirit that wishes to approach the stones' state, but at the same time experiences joy in meditating upon them.

VI

Critics of his writings about stones remark on a whole range of conflicts that Caillois manifests rather than overcomes.[48] In the first place the stone itself is unmoving and completely present, whereas the text that describes it develops in time, and gradually introduces movement into the stone.[49] On the level of ideas about the stone a whole range of contradictions is established, and these remind us of archaic speculations about the elements—hot and cold, dry and moist, solid and liquid, dark and light.[50] The stone is solid and unmoving, but at the same time is the result of a balance of forces that emerged from burning chaos;[51] originally this was liquid molten matter which cooled and hardened. In several cases Caillois also describes stones as an image of a cold and inhospitable waste land, which contrasts with their original molten state. Stones are a fragmentary image of the world. Through their shapes or the drawings on them, they remind us of the natural order, and at the same time seem to deny the existence of such shaping and order, since resemblance proves deceptive. As objects of meditation, they possess special depths that give them priority over other created objects. They provide space for the great human need to search out the familiar in random groupings, even as they entirely deny the possibility of such resemblances.[52]

---

48 Starobinski characterizes Caillois generally by two co-existent poles of thought: dualism and the demand for unification. "Saturne au ciel des pierres", 185.

49 Rigoli writes of the paradoxes of "writing on stones" (*L'Écriture des pierres*, as one of Caillois's titles has it), and the discrepancies of stone and discourse. "Caillois, le lapidaire intime," 425.

50 Stone as the trace of previous processes embodies these paradoxes: it is cooled lava, solidified liquid etc.

51 *Œuvres*, 1046.

52 In *Esthétique généralisée* (A General Aesthetic, 1962) Caillois attempted to formulate a more general aesthetic theory, summarizing both human art and natural phenomena. In his writings on stones his concern lies somewhat elsewhere, as though the stones call human art into question. For instance, cf., "Autre cuivre" (Another Copper) which compares a piece of natural copper with a fragment from a work of sculpture (*Oeuvres*, 1049–1051).

We can also read Caillois's meditations on stones as a sceptical consideration of the human in the second half of the twentieth century. Humanity finds nothing that would confirm a given order of things, as though the world itself might be shapeless, as Caillois's contemporary Bataille supposes. To a god-like observer who could view it from without, our universe would appear no more than a gobbet of spit, no matter what relative arrangements are discovered by humans locked inside this shapelessness.

Stones embody this conflict between the shape and shapelessness of the world, even while in certain situations they can become fixed points. They are not alive, and never were. No part of them is *functional* as, say, part of a plant or living organism. They do not signify. If the observer searches for some meaning in them, he or she finds only the traces of the past. Stones proffer a particular silence and an emptiness of non-communication and unchangingness into which a human being can immerse him- or herself through thought and feeling. In these depths, the stone comes forth as a trace of a prior state that preceded humankind. It always passively cedes to human challenges, turning to its inception—an inception that is also imaginary, since it is a dreamed extrapolation of the traces in the stone. In contrast with living humans, the time of stones is revealed as a state that persists, having experienced all change in the inaccessible past.[53] This temporal "passivity" reminds us of the brevity of human life, and thus becomes the point at which the human is forced to turn back upon itself. Thirdly, this passivity is evident in the fact that it resists language. Every attempt to describe the material presence of stones (which omits their genesis) finally comes up against something ineffable and swerves into analogy. This resistance to language points up the hiddenness of stones, as though they are only partially manifest on the level of language. For thought, in which depth and dreaming are joined, stones cease to be objects and become such a turning point; a thought which arrives at a stone can go no further, and necessarily turns back, as though it were the beginning, or inception. The stone in its passivity is manifest as an object drawn violently into the field of the visible, and it continues to resist that violence.

Translated by Justin Quinn.

References

Bachelard, Gaston. *La poétique de l'espace*. Paris: Presses universitaires de France, 1998.

53 On this temporal precedence, cf. Massonet, *Les Labyrinthes de l'imaginaire*, 277.

— *The Poetics of Space*. Translated by Maria Jolas. Boston: Beacon Press, 1994.
— *La poétique de la rêverie*. Paris: Presses Universitaires de France, 1989.
— *The Poetics of Reverie: Childhood, Language and the Cosmos*. Translated by Daniel Russell. Boston: Beacon Press, 1969.
— *La terre et les rêveries du repos*. Paris: Librairie José Corti, 1997.
Caillois, Roger. "Aventure de la poésie moderne". In *Approches de la poésie*, 51–64. Paris: Gallimard, 1978.
— *Œuvres*. Paris: Gallimard, 2008.
Massonet, Stéphane. *Les labyrinthes de l'imaginaire: dans l'œuvre de Roger Caillois*. Paris: L'Harmattan, 1998.
Rigoli Juan. "Caillois, le lapidaire intime." In *Mémoire et oubli dans le lyrisme européen. Hommage à John E. Jackson*, edited by Dagmar Wieser, and Patrick Labarthe, 409–436. Paris: Honoré Champion, 2008.
Starobinski, Jean. "Saturne an ciel des pierres." *La Nouvelle Revue Française* 320 (1979): 176–191.

—

Josef Hrdlička
Department of Czech and Comaparative Literature
Faculty of Arts
Charles University in Prague
Jan Palach Square 2
116 38 Prague 1
Czech Republic
josef.hrdlicka@gmail.com

www.ingramcontent.com/pod-product-compliance
Ingram Content Group UK Ltd.
Pitfield, Milton Keynes, MK11 3LW, UK
UKHW041828100926
8556IPUK00002B/75

9 783631 650691